Watershed Development and Health

Study of Child Nutrition in Rural Semi-Arid Region

Apoorvaa Pandit

Dipak Zade

WATERSHED ORGANISATION TRUST (WOTR)

Pune, Maharashtra, India

RAWAT PUBLICATIONS

Jaipur • New Delhi • Bangalore • Hyderabad • Guwahati

ISBN 978-81-316-0544-8

Published by
Prem Rawat for **Rawat Publications**
Satyam Apts, Sector 3, Jawahar Nagar, Jaipur 302 004 (India)
Phone: 0141 265 1748/7006 Fax: 0141 265 1748
E-mail: info@rawatbooks.com
Website: www.rawatbooks.com

New Delhi Office
4858/24, Ansari Road, Daryaganj, New Delhi 110 002
Phone: 011 2326 3290

Also at *Bangalore*, *Hyderabad* and *Guwahati*

Typeset by Rawat Computers, Jaipur
Printed at Chaman Enterprises, New Delhi

Contents

Foreword

The recent statement by Prime Minister Dr Manmohan Singh that 70 per cent children in India are malnourished, has sparked off an intense debate on the issue. In fact, infants and children of today are our valuable human resources of the future. The way we nurture them today, will determine the quality of our human resources tomorrow. Child nutrition and health is a universal goal of development.

The social, economic and educational development has a direct impact on human health and nutrition. Development of a country in various sectors, such as agriculture and industry, helps not only in meeting the needs of poor people but also improves access and availability to various resources. Development not just means increase in a person's or family's income, but it reflects the status of availability of clean drinking water, sanitation, good nutrition and also the education and status of women in the society.

Malnutrition is higher in all developing countries. India is no exception with a higher incidence of childhood malnutrition, as indicated by National Family Health Survey. In the semi-arid and drought-prone rural areas water is the key to progress and development. A recent study by the authors indicates a change in the nutrition and health consciousness of the people where watershed development and child growth inputs are given. The present work extends this understanding by a detailed inquiry into the nutritional status of children aged 0–5 years as against the developments that have taken place in the villages.

Height-for-age is an impact indicator of watershed development. The study has formulated a hypothesis expressing the critical role of water and soil conservation measures in the context of child nutrition, which is based on sound observations. The study is vast. The village group with watershed development and child growth inputs has been compared with two others: one, where only watershed development has taken place with no child growth inputs and the other where no such kind of inputs are given. The results indicate nutritional status as slightly better in control group so far as height-for-age index is concerned.

Hypothesis may or may not get proved. What is important is the knowledge generated that explains the role and interplay of different factors that contribute to the process and the present study has very well put forward the explanations.

The study has used a variety of tools: anthropometry, clinical assessment, diet survey and food frequency to understand the nutritional status of children in depth. The social, economic and developmental status of the different villages has been presented at length; the profile of the mothers of the children under study has been generated and discussed, as these are important factors affecting child nutrition.

The subject matter is complex. Most agriculture intervention increases food production, but does not necessarily improve nutrition or health within the participating households. It is understood that a complex relationship exists between production, income and nutrition. The researchers have competently handled the huge data, the differences and certain paradoxical situations in the results and the comparative analysis that posed a great challenge in the interpretation of the results. These have been duly explained with qualitative analysis.

The book is extremely useful to students of social sciences and to professionals from different sectors as well. It adds a new dimension to research: of health to watershed development and that of watershed development to health.

Shaunak S. Kulkarni
Associate Professor, Department of Anthropology
University of Pune
Pune

Acknowledgements

The present study has been benefited greatly by support and cooperation from many individuals and we would like to thank them all.

At the onset, we would like to thank Dr Marcella D'souza, Executive Director, Watershed Organisation Trust and our Research Advisor, for her support and guidance. She has been a patient listener to the problems we faced at varied levels in the study. She has gone through the entire manuscript of the book and has also made valuable suggestions.

We acknowledge with much gratitude the timely inputs of our Research Advisors, Dr Shaunak Kulkarni, Associate Professor, Department of Anthropology, University of Pune and Dr Chandrakala Mannuru, Reader and Head, Department of Food Science and Nutrition, S.N.D.T. College, Pune.

Many thanks are due to Dr Swati Tongaokar, Clinician (Consultant), for conducting clinical assessment of the children in different villages under study and to Dr B.V. Bhanu, Professor

(Retd.), Department of Anthropology, University of Pune for inputs in the writing of the chapter, "Water, Development and Child Nutrition".

Our special thanks are also due to senior officials of WOTR and Sampada Trust, Ahmednagar. Mr Ganesh Rajapure in selection of the study villages and Mr Joseph Shinde in selection and nomination of students of WOTR as Field Investigators for the study. Ms Chayya Nikrad provided data from WOTR database, Ms Preetilata Gaikwad provided field inputs and Ms Ujjwala Kavthekar and Ms Bhagyashree Shinde provided field assistance whenever required.

We would also like to express our sincere thanks to Mr Kishor Telang and Mr Ajay Shelke, Pune, Mr Date and staff, Darewadi Training Centre, Darewadi, Ahmednagar for their administrative support.

Many thanks to Ms Aparna Ghodke and Ms Anuprita Dixit, Statistical Consultants, who conducted data entry and analysis for the study. Thanks to Mr Sanjay Kulkarni for carrying out the translation of interview schedules and training material.

We would also like to acknowledge the help of our colleagues: Mr Atul Pagar for providing the photographs, Mr Santosh Ramane for computer-related assistance and Ms Aparna Chaudhari, for helping us in the field. Ms Pradnya Ghag assisted in preparing the maps of the study villages.

Thanks to Ms Sindhu Dhondge, Ms Savita Baris, Mr Swapnil Bhilkar, Mr Navnath Ighe and Mr Kailash Lokhande who have carried out the data collection work as Field Investigators.

We wish to express our gratitude to the *anganwadi* workers of the study villages for their cooperation and support during the field work. They were always willing to help us. Last but not the least, we wish to thank all the people from study villages for their hospitality, accommodation arrangements and for sparing time to answering our questions. We wish to specially thank Ms Nandatai Kale and Ms Sangeeta Bodke, Health Promoters and Mr Dilip Satkar from village Mhaswandi, Ahmednagar in this regard.

Lists of Tables, Figures and Plates

Tables

Figures

Plates

1

Introduction

Development: A Backdrop

"Development" is a highly complex and multifarious process. The perspective on what constitutes development may differ from culture to culture because it depends on how a community perceives it. Besides, there could be regional and gender differences in the perception of development. It has social, cultural, economic, political, environmental and health dimensions and implications. Literature cites that the goals of development are associated with poverty alleviation, livelihood security, quality of life, cultural and political freedom, increased well-being, gender equity, effective public services, sustainability of environment and socio-economic growth. On the whole, it could be said that economic growth, social advancement, health improvement and environmental sustainability could lead to better opportunities for development. Health and nutrition is a cause and effect of the goals associated with development. Whether human development is achieved or not as desired, there is a definite growth in the understanding of the concept of development. The concept changes with the passage of time. The meaning of development has changed from economic growth to equity towards sustainability during the period from 1960 to 1990 (Levinson and Melvin, 1996).

Development means alleviation of poverty. The concept of poverty too has many facets. There is also a correlation between poverty and child nutrition. A glimpse of poverty in India shows that child malnutrition has affected almost half of the population. The terms as described above set as goals of human development are broadly overlapping and imply more or less the same meaning. For example, alleviation of poverty means greater livelihood security, enhanced quality of life, having lesser children with malnutrition. Poverty cannot be alleviated without development in health and education. Quality of life is dependent on the availability, accessibility and effectiveness of public services. It includes access to health care, education, employment opportunities, and availability of safe drinking water. Literature points out that poverty also means a state of powerlessness (weak negotiating power), oppression and injustice. Empowerment is a sign of development. Empowerment of the disadvantaged and the weaker sections of the society, for example, the women, is particularly emphasised. Women development in turn leads to child development.

Development is meaningless without the sustainability of the goals achieved. Sustainability implies equilibrium and a thought for future generations. Development is sustainable if it meets the needs of the present generation without compromising the ability of future generations to meet their own needs. It is equitable and balanced, meaning that in order for development to continue indefinitely, it should balance the interests of different groups of people, within the same generation and among generations, and do so simultaneously in three major interrelated areas (Carney, 1998). Development and sustainability are ongoing processes of change. Child nutrition is a mirror of sustainability of the development achieved.

"Capacity building" is the term often associated with development. Planned development actions focus on strengthening the capacities of the people in two major ways: (i) transfer of technology, and (ii) skills and knowledge. The concept of development, the problems of development as well as the efforts for planned development seem to vary from region to region. For example, water-scarced arid regions of India are most susceptible to degradation of natural resources and recurrent droughts, which have been a cause for impoverishment of the inhabitant population (Chauhan, 2004). Impoverishment is associated with low productivity of human and livestock resources. In such regions, water and soil conservation measures, water harvesting techniques and watershed development are the major inputs for the development of the arid, semi-arid and rainfed regions.

Enhanced capabilities with enhancements in knowledge and skills are considered as signs of development. The socio-economic life of a particular area is greatly influenced by its dynamic interactions with the various development organisations, government, non-government or other bodies who initiate to bring about a planned change. The status of child nutrition exemplifies the capacity of the people developed for adopting the strategies to improve the nutritional status of children.

As development has many facets, it could be studied from many different perspectives and by using different indicators.

Development takes different forms in different types of regions. For example, the Green Revolution was introduced in the areas having assured water supplies. It increased the efficiency of agricultural processes so that the productivity of crops was increased. Food production increased substantially. It also introduced regional inequalities and income disparities with newer social problems arising out of these.

Watershed Development

Watershed development is introduced in rainfed, semi-arid and arid regions to create sustainable livelihood opportunities for the rural poor. Most of the small farmers are living in arid and semi-arid areas. Watershed development is a revolution of another kind. By means of efficient water and soil conservation techniques, water management systems, and sustainable use of natural resources, it brings about agriculture and livestock development among the small farmers living in these areas. Water management is a pre-requisite for livelihood today especially where ground water levels are depleting at a faster rate and water crisis is looming. Availability and access to water for drinking and for agriculture, a primary means of food security and livelihood, is dependent on efficient water management. Watershed development is a technique of water management that harnesses rain water and stimulates agricultural growth in rainfed, arid and semi-arid regions. Evidences show that watershed development has led to economic prosperity in these areas by means of agriculture and livestock development.

A 'watershed' can be defined as the drainage basin of a catchment area of a particular stream or river including glaciers. In simple words, it refers to the area above any point on a defined drainage channel, which feeds water into it. It may vary from a few hectares to several thousands of hectares.

Watershed development refers to the conservation, regeneration and judicious utilisation of all the resources: land, water, vegetation, animals and humans within a particular watershed. Watershed development seeks to bring about an optimal equilibrium in the eco-space between natural resources, man and animals.

In Ahmednagar, Indo-German Watershed Development Programme (IGWDP) has been implemented in several villages. It is a comprehensive programme based on the principles of integration, community participation, self-help and environment regeneration. It is carried out through the formation of an organisation, Watershed Organisation Trust (WOTR), which is a non-government organisation (NGO), in the year 1993, based in Ahmednagar. WOTR is the official capacity building organisation of IGWDP. It is the resource organisation in the field of watershed development. The programme involves the integration of the following components:

(i) Human resource development (community development)
(ii) Soil and land management
(iii) Water management
(iv) Afforestation

(v) Pasture development
(vi) Agriculture development
(vii) Livestock management

The watershed development programme as implemented through the approach of WOTR is divided in two parts (WOTR, 2005).

Capacity Building Phase (CBP)

The purpose of the CBP is to enhance the capacities of the villagers to implement the watershed development programme effectively. Capacity building takes place by the formation of various local level institutions in the villages such as the village watershed committee also called as Village Development Committee (VWC/VDC), women's self-help groups (SHGs), women committees, or the *samyukta mahila samiti*[1] (SMS), which is the apex body of the SHGs in the village (WOTR, 2005). The CBP instils development by enhancing the values of self-help and team spirit. A micro-watershed of about 100–150 hectares is undertaken on experimental basis as an activity for demonstrating and learning process. This is done through *shramdan*, i.e., through voluntary contribution of labour. The local institutions undergo orientation programme to understand the concept of watershed development. The village people are oriented towards various physical and technical aspects of watershed development. They also undergo training schedule to equip them with the social, technical and managerial skills required (NABARD, year not mentioned). This phase lasts for about one to one-and-a-half year and is implemented to judge the commitment and capacities of the villages in implementing the watershed development in the village.

Full Implementation Phase (FIP)

During the FIP, physical watershed development work is carried out on the land through various land and soil treatments. This phase lasts for three to four years depending on the size of the watershed. Simultaneously, the strengthening of the local institutions and women's promotional activities are carried out. By this time, the local institutions become capable of taking their own responsibilities.

The watershed development activity is carried out in a ridge to valley basis. The land on the hill and the hill slopes, which are at higher elevation, are treated first followed by the low-lying areas. This approach has many benefits. It ensures maximum water conservation and ground water recharge. It also helps in effective reduction of soil erosion. Socially, this approach is also beneficial to the poor and the marginalised sections of the village. Generally, this group owes land on the hilly terrain and slopes and the affluent people have land at low lying areas or at the valley. This treatment of the land on the hilltop done on priority basis ensures that the benefits of the watershed development first reach the poor and the marginalised group of the society.

Watershed development is based on three important principles. They are: bans on felling of trees, grazing of cattle in open areas and voluntary contribution of labour to implement the programme in the village.

Sampada Trust

Founded in 2002, it is a sister organisation of WOTR and undertakes a variety of activities like women empowerment, micro-finance, micro-enterprise development and micro-insurance, etc. It also provides women empowerment services by formation of SHGs of the poor women, their capacity building, training, extension micro-finance and insurance facility to the rural poor women across eight districts in Maharashtra. Some of its activities are described as under:

Social Capital Formation: Promotion of Women's Self-Help Groups

Sampada Trust organises women into SHGs and helps in its promotion in the village. These SHGs are federated in an apex body at the village level called as Samyukta Mahila Samiti (SMS). Two members from each of the SHGs are represented in the SMS. At the initial stage, the SHGs deal with credit- and savings-related activities and at a later stage, they are involved in social development activities in the village, particularly in the field of education, health, sanitation and hygiene, etc. These activities also facilitate mainstreaming of the women in decision-making process in the village.

Capacity Building through Trainings

Training exchanges and exposures on relevant themes are organised regularly for the SHGs, SMS, Mahila Pravartak (Women Health Promoters), NGOs and its staff. These activities help in capacity building of these organisations as well as individuals by increasing knowledge and information.

Some of the important programmes deal with issues such as, Panchayat Raj Institutions (PRIs), management of SHGs and SMS, their records and documentation process, management of micro-finance and micro-enterprises for NGOs, leadership skills for SHGs and NGO leaders.

Sampada Trust organises trainings to enhance the entrepreneurial competencies and managerial capacities of women, unemployed youth, NGOs, CBOs, etc. The trainings are organised as per the need, viz., capacity building, specialised training and skill-based training.

Micro-Credit

Sampada Trust also provides credit to the SHGs directly or through SMS. This way micro-finance reaches the poor who are outside the formal credit system. They are also provided the necessary managerial and technical support for starting their own micro-enterprises. Such support, along with the micro-credit, helps them earn higher income and come out of poverty.

Micro-Enterprise

Sampada Trust motivates and provides technical information to the SHGs through the SMS to undertake income-generating activities and micro-enterprises.

The potential for such activities, resources and women willing to undertake such activities are identified initially. The women are then provided training, covering aspects of motivation, financial organisation, accounting and marketing management. Along with providing the credit, the women are also provided inputs to make their enterprises sustainable and economically viable. Sampada Trust has established a retail urban outlet, Amhi Leki Ahilechya, for the sale of the products of rural SHGs and artisans at Ahmednagar.

Social Security: Micro-Insurance

There is an urgent need to secure the lives and livelihood of the rural women. However, the insurance services are not accessible for the rural poor. Considering this need of women, Sampada Trust initiated provision of insurance policies at their doorsteps. Sampada functions as a corporate agent to link rural individuals to reputed insurance providers and insure their life and accidental death.

Social Development Activities

Sampada Trust is also involved in social development activities, such as drinking water, latrines and sanitation, health and hygiene promotions, exchanges and *melawas*, etc. These activities have made the women confident. *Melawas* or gatherings of women are the means to build their capacities to come to the forefront and share their practical experiences of implementation of their own thoughts and dreams. A *melawa* provides a platform to the rural women to build networking of relationships, bonds of solidarity, and exchange experiences from different villages and empower women with the required knowledge. During the *melawas*, the participants are provided inputs regarding health, social issues, legislation, various enterprises and approaches towards empowerment.

Social and drudgery reduction activities like smokeless *chulhahs*, drinking water systems, kitchen gardens, LPG, toilets, etc., have been taken up through these groups. Besides this, income-generating projects like flour mill and canteen run by women, etc., have also been taken up. Sampada Trust has also provided solar home light systems in many of the villages.

Child Care and Growth Monitoring Activity

This activity is being implemented in the villages to improve the nutritional status of children in the 0–5 years of age group. It includes creating awareness among women regarding the importance of balanced diet in coordination with the local Anganwadi programme of the State Integrated Child Development Scheme (ICDS). Activities such as regular height and weight check up, demonstration of balanced diet, healthy baby competition and immunisation in coordination with the Primary Health Centres (PHCs), Anganwadi Sevika Workshop, trainings for kindergarten helpers, and melawas (training and outreach social gatherings) for parents and children are carried out.

The SHG members are motivated to grow kitchen gardens in their households. They are provided inputs regarding types of vegetables to be grown in these kitchen gardens which are healthy and have nutrient-rich contents. These vegetables could be used to make nutrition-rich food items without spending the money for purchasing them from the market.

Women health volunteers, the *Mahila Pravartaks,* have been trained for taking body measurement like height and weight of children aged 0–5 years. The growth of children is monitored once in a month. The mothers of the children are involved in plotting the weight of their children on the growth charts developed by the organisation. This helps them in understanding the nutritional status of their children. The health volunteers guide the mothers to adopt strategies to improve the nutritional status of their children. The activity is implemented in coordination with the *anganwadi*[2] workers (www.sampada.org).

As there are huge investments made in watershed development programmes, in terms of skills, capacity building, technical and managerial support, funds, time and labour and other resources, it becomes imperative to assess the effectiveness of such programmes. Nutritional status of children (height-for-age) aged 0–5 years is one of the indicators recommended to measure the impact of water and soil conservation (Bollom, 1998).

Plate 1: Parents and children meeting with health promoters

Plate 2: Food demonstration by health promoters in watershed and growth monitoring village (WSD+GM group)

Watershed Development and Health

Nutrition is an important aspect of health and human development. Child malnutrition is considered as a key indicator to describe human development. It has a series of negative consequences in several areas. Most crucially, malnutrition has an impact on morbidity and mortality, and also on education and productivity, thus constituting one of the major mechanisms for the inter-generational transmission of poverty and inequality (www.eclac.org).

Malnutrition is a worldwide problem. The prevalence is higher in developing countries and particularly in Asia. A higher level of undernutrition in India is well recognised.

Malnutrition is frequently part of a vicious cycle that includes poverty and disease. The three factors, viz., malnutrition, poverty and disease, are interlinked in such a way that each contributes to the presence and permanence of the others. While malnutrition manifests its diverse forms, protein energy malnutrition ranks foremost in terms of socio-economic consequences and the enormity of its effects (Kaur et al., 2005).

In the rainfed regions characterised by fragile resources and land degradation, child nutrition is both a cause and an effect of watershed development achieved. Low productivity of agriculture and livestock in these regions contributes to malnutrition in a major way. There are several factors, which contribute to the growth of agriculture and livestock sectors and in turn reduction in poverty and child malnutrition. Water is the basis of all. Availability and access to water resources plays a critical role in improving agricultural yield and returns from livestock. Natural resources, farm productivity, access to markets, appropriate infrastructure, newer technologies, socio-economic conditions, culture of the area and policies are some of the other factors.

Watershed development plays an important role in enhancing the water resources in rainfed regions subsequently enhancing human capabilities to improve upon and achieve the overall development of a region that contributes to child nutrition. It is said that watershed development ensures greater food and livelihood security due to increase in land capability, agricultural yield and allied income opportunities and by arresting the process of distress migration. Increase in agriculture production and productivity is the major impact of the programme. Soil and water conservation work improves the soil moisture regime and increases the water level in wells, which helps the farmer in cultivating more crops than before.

Availability of regular labour opportunities through watershed work reduces distress migration to a large extent.

Watershed development leads to an increase in water table causing changes in other areas leading to development. A study conducted by WOTR, to document the impact of watershed development in different villages in Maharashtra, shows an increase in water table. A list of indicators was used to measure the change. With the rise in water table in the villages an increase in the number of the indicators was observed as shown below:

Table 1.1

Comparison of pre- and post-watershed development situation

Impact indicators	*Number of villages*	*Pre-watershed development*	*Post-watershed development*	*Per cent increase*
Wells (number)	23	2,130	2,684	26
Agriculture assets* (number)	17	1104	1694	53.4
Total irrigated area (hectare)	23	2,074.3	5,306.5	155.8
Cropped area: Rabi (hectare)	21	5,327.2	6,960.8	30.7
Cropped area: Summer (hectare)	11	76.4	465.6	509.6
Agriculture production** (quintal per hectare)	22	67758.8	107608.1	58.8
Fodder availability (tonnes per year)	21	22297.4	70707.3	217.1
Cross bred cows (number)	28	1923	3174	65.1
Milk production (average litre per day)	17	203.9	777.1	281.1
Individual latrine (number)	21	149	909	510.1
Cement houses (number)	9	158	430	172.2

Notes: *Agriculture assets include tractor, electric motor and oil engine.
** Agriculture production includes production of cereals and pulses.

Source: Unpublished WOTR data.

Water and soil conservation efforts boosted following developments in the villages:

1. Availability of safe drinking water and hygienic sanitation practices.

2. Multiple cropping and higher productivity.
3. Dairy development.
4. Increased community participation. People's institutions and women's participation in development process are examples.

A question that raised during these observations was: "What is the status of nutrition of children aged 0–5 years in these kinds of villages and with this kind of development at the background?"

A qualitative, comparative study conducted in three different villages in the semi-arid regions of Maharashtra indicated the changing culture of child nutrition. It exhibited a change in the social and economic parameters, a change among the women, increased child health and nutrition awareness and related behaviour in different communities and other different groups of people (Pandit, 2010). However, the question still remained unanswered and that was whether all these changes have influenced the nutritional status of children aged 0–5 years, as indicated by anthropometry, particularly the height and weight. Bollom (1998) indicated that the role of water and soil conservation measures is critical in nutrition and that improved level of nutrition, particularly the height-for-age is one of the indicators of water and soil conservation.

Very few studies have been undertaken to assess the impact of water and soil conservation measures in terms of nutritional status of a population in general and in the context of dry, semi-arid and drought-prone areas of Maharashtra in India. We therefore decided to take up a study to understand the nutritional status of children aged 0–5 years by using anthropometry as against the social and economic profile of the families of these children and their villages. As there are not so many studies which talk about the impacts of watershed development in terms of nutritional status of children aged 0–5 years and which would help to form a base to develop the study that we had in mind, we decided to learn by developing this kind of study. The present work should be considered as the pilot study in the field. Initially we conducted a small study based on clinical method. Its methodology, results and experiences are shared in the next chapter.

The present work discusses the results of a study that has been undertaken to evaluate the impacts of watershed development and participatory growth monitoring[3] inputs. It discusses child nutrition in the context of development of villages. There are different groups of villages: one, where integrated watershed development and child growth inputs have been initiated as compared with a group having a single input of integrated watershed development and that with none of these inputs. It discusses the analysis of the nutritional status of children aged 0–5 years and the status and form of development that has taken place in the rainfed and semi-arid regions of western Maharashtra. Height-for-age is used as an indicator to measure the impact of watershed development. The work is based on a study conducted in the rainfed rural areas of Ahmednagar. Ahmednagar is a district in western Maharashtra. It is a dry land area and resource-poor region. The area on the whole is characterised by low soil fertility, low and erratic rainfall, rugged terrain, remoteness and poor infrastructure.

Notes

1. The *Samyukta Mahila Samiti* is the apex body of Self-Help Groups (SHGs). It is a village level federation of the various SHGs.
2. Anganwadi is a child care centre located within the village through which Government Integrated Child Development Services are provided.
3. Participatory growth monitoring is conducted by Sampada Trust, an NGO in Ahmednagar and sister organisation of WOTR. The activity includes monthly monitoring of weight and height of children in the villages. The women who voluntarily decide and those who are nominated by village people are trained as health promoters *(Mahila Pravartak)* by Sampada Trust to undertake monthly growth monitoring of children through home visits. The mothers of the children are involved in plotting the weight of their children, which helps them to understand their children's nutritional status. Nutrition education is given to the community through balanced diet demonstrations in the villages. Besides, the health promoters guide the mothers to improve the nutritional status of their children through a variety of supplementary food.

2

Water, Development and Child Nutrition

This chapter is presented in the form of a paper developed out of a small study conducted alongside the intervention activity. The paper reports the nutritional status of children from the rural dry belt where long-term efforts of increasing water levels through water and soil conservation measures and other developments have taken place.

Child nutrition is an indicator of successful water and soil conservation. Evaluation of ecological and economic impact of watershed programmes has been the subject of concern since long. Clinical method helps to rapidly analyse the status of child nutrition and assess the impact of planned change through watershed development programmes. Huge investments and efforts are put into programmes for watershed development and child nutrition separately in India. Child nutrition is a result of years long water-related and corresponding social development. An integration of watershed and child nutrition programmes would enhance further progress, through active involvement of the people's institutions.

Child nutrition and health is a universal goal of development. Nutrition in the early stages of life has long-term effects on productivity in adulthood. Malnutrition arising from inadequate diet, micronutrient deficiencies and infections contributes to morbidity and mortality in childhood affecting health and productivity. Malnutrition is higher in all developing countries. India is listed as one of the countries that have yet to achieve the target of the World Food Summit of 1996, in terms of reducing the number of undernourished (National Academy of Agricultural Sciences, 2003). It is also one of the countries with a higher incidence of childhood malnutrition as indicated by National Family Health Survey (NFHS) figures. NFHS-3 states that almost half the number of children under five years of age (48%) are stunted and 43 per cent are underweight (IIPS-Macro, India Report 2007).

There is a strong relationship between water, development and nutrition. One of the most important means to development is water. Water is basic to life. It serves many essential purposes of life. Drinking, the most important survival need, cooking, bathing, personal hygiene, environmental sanitation, domestic purposes like washing clothes, cleaning the house and utensils, homestead cultivation, agriculture and other income-generating activities are all dependent on water. Water is the key to economic and social development. It generates as well as provides a means for diversification of livelihood. Availability of adequate water is a prerequisite for the survival and maintenance of life. The quality of water (safe drinking water) is as important as its quantity as it is related to the health of the people. The issue of access to adequate as well as safe drinking water is more pertinent to the rural areas. In the semi-arid and drought-prone rural areas progress and development are built around water. A report by the Government of India (Subramanian, 2002) states that in the six decades since independence, phenomenal development of water resources has taken place in India. It is claimed that availability of drinking water for about 85 per cent of India's rural and urban population has been achieved. This largely contributes to self-sufficiency in food grains. Besides, there has been rapid expansion of irrigation and drainage infrastructure in India. The irrigated agriculture contributed to the country's economic development and poverty alleviation.

The water and sanitation sector has received considerable attention since the International Drinking Water Supply and Sanitation Decade (IDWSSD) 1981–1990 (Gopalan and Saksena, 1999). A revised National Water Policy, adopted in the year 2002, accorded primacy to drinking water. Access to adequate as well as safe drinking water and sanitation is a precondition for child health (Lenton et al., 2005).

Agriculture and livestock are the primary means of livelihood of the rural poor. Most of the agriculture in India is rainfed. In Maharashtra alone, more than 80 per cent of agriculture is rain-dependent. Therefore, the rural rainfed regions of India and Maharashtra reveal a grim picture of water scarcity, rapid depletion of ground water table, fragile ecosystems and poverty. In the rainfed semi-arid regions characterised by low productivity and degraded natural resources, water and soil conservation is of utmost importance to enhance productivity, income and livelihood security in a sustainable manner. Watershed development is an efficient means of water and soil conservation. It involves water harvesting and efficient water management that contributes to sustainable development of rainfed areas.

Evidence indicates that watershed development leads to economic prosperity and reduction in poverty by means of agriculture development, livestock extension and environmental regeneration in the long term. A number of studies indicate improvements in food security through increased crop yield, higher milk production, increased livelihood opportunities, employment generation, better asset position and reverse migration. These are recognised as social and economic changes that have come about after watershed development in the villages. It is hypothesised that due to watershed development there is an increase in the variety of food components and quantity at the household level, and that the daily consumption of milk increases due to livestock extensions contributing to better nutritional status of children.

Dietary Deficiencies: Malnutrition Problems in Brief

Nutritional deficiencies, especially as observed in Asia and Africa, are a result of extreme poverty because poverty limits diversity in the diet. Poor nutrition is manifested in two forms. One is nutritional deficiency which is not easily visible (hidden hunger). Another is malnutrition that is visible and measurable (visible hunger) (Zeigler, 2006).

Malnutrition is a pathological state resulting from a deficiency or excess of one or more essential nutrients. Protein Energy Malnutrition (PEM) is the most common form of malnutrition occurring in India. It is caused due to inadequate amount of nutrients, both in quality and quantity such as proteins, fats and carbohydrates in the diet. It usually occurs in children aged 0–5 years. It leads to childhood morbidity and mortality, and permanent impairment of physical and mental growth. Children with PEM are highly susceptible to infections, for example, diarrhoea and anaemia. There are many other problems arising out of malnutrition. Anaemia is caused due to the deficiency of iron in the diet. The children are weak and show pallor of the skin. Also, Vitamin-A deficiency is a major public health problem in India. Night blindness is one of the main clinical conditions caused due to Vitamin-A deficiency. Vitamin-D deficiency causes rickets in children. It is characterised by deformities in bones. The ankles, knees, wrists and elbows are swollen (MSBSHSE, 2007 and Nutrition Society of India, 2005).

The Study

A child health intervention and study activity was conducted as a part of a large-scale research study on child nutrition carried out at WOTR, Pune.

The purpose of the intervention was to provide general health check-up for children aged 0–5 years and provide medical advice to the community for improvement of nutritional and health status of their children, thereby generating community awareness on child nutrition.

Health camps were organised to conduct the intervention.

Objectives

The objectives of this study and intervention activity were as follows:

1. To detect PEM and nutritional deficiencies in children aged 0–5 years.
2. To generate a database of child nutrition for the organisations working towards watershed development for long.
3. To give feedback to the community about the nutritional status of children after having conducted a study in that area.
4. To identify needs for undertaking intervention in the area of child nutrition strengthening water development programmes.

Data Collection Method and Procedure

The data were collected by means of clinical examination of children by a physician through interviews with the mothers and/or care givers of children who accompanied the child in the health camp. The investigators having a disciplinary background in Anthropology and Social Work were also a part of the study team. Anthropologists were involved in in-depth interviews with different people, conducting observations, note taking, helping the doctor and establishing rapport with the people. The social worker team member assisted in planning, organising and in management of health camps. Informal in-depth interviews were held with key informants of the village, like village level workers, *anganwadi* (ICDS/Integrated Child Development Services) workers and elderly persons and also observations were conducted. Two types of observations were made. One regarding practices related to child care during the health camp. The other related to the observations in the villages. The data are collected with the help of interview schedules. The interview schedules are of two types: one designed for clinical examination and the other for understanding village development profile. The clinical examination schedule is based on the clinical signs detecting nutritional deficiencies.

Clinical examination is one of the simplest and most practical methods of assessing nutritional status. There are a number of physical signs, some specific and many non-specific, known to be associated with malnutrition. The method involves looking for changes in the body, which are indicative of a particular deficiency. There are many advantages of this method; e.g., no elaborate field equipments and costly laboratory procedures are needed (Nutrition Society of India, 2005). The method is of great use to development organisations where intervention and research activities are carried out. It helps to rapidly assess the impacts of development and plan future interventions. Because of the simplicity of this method community health workers can also be trained to recognise certain clinical signs.

The clinical examination study gathered information on the following components:

(i) Background information of the child
(ii) General clinical features
(iii) Clinical features of PEM
(iv) Nutritional deficiencies

The study focused on understanding the physical and mental growth, prevalence of PEM, anaemia, nutritional deficiencies related to Vitamins A, D and that of calcium, widespread skin infections, worms, diarrhoea, dental problems and any other illnesses.

Some secondary data have been collected as a supplement to the primary data. These data have been collected from *anganwadi* records from each village regarding nutritional status of children and that of child mortality. The *anganwadi* focuses on children aged 0–6 years and their records of nutritional status of children are based on weight-for-age criteria. The weights of children are measured by the *anganwadi* workers with the help of spring balance weighing method. The weights are plotted on ICDS growth charts and the status is recorded as normal or in malnutrition grades from I to IV. Other secondary data were collected from WOTR database to understand the changes occurred in villages after watershed development.

Sample Size

As the study was a part of intervention activity, the sample size was not predetermined. Health camps were meant for all the children in the villages selected and participation was voluntary. The response to the health camps was overwhelming. A total of 174 children aged 0–5 years were covered in clinical examination. The villages where a previous study had been carried out were selected for the present health intervention and study so as to give feedback to the people.

Study Site

The clinical study was conducted in five villages: Mhaswandi, Gunjalwadi, Karjule Pathar, Kangar and Kalas Pimpri in Ahmednagar district of Maharashtra. The district is a semi-arid and drought-prone one. These are also the villages where water and soil conservation has taken place. Except village Kalas Pimpri, in all other villages, integrated and participatory watershed development programme has been undertaken. WOTR has played a leading role in providing technical support to the local organisations implementing watershed development work. The villages Mhaswandi and Gunjalwadi have additional benefit of a development input. In these villages NGO initiated participatory growth monitoring of children.

Health Camps

The health camps were held during the period from February to May 2008. One health camp each was held in all the five villages. The dates, venue and schedule of health camps were fixed in consultation with the village people as per their convenience one week prior to the camp. A meeting was held with the key informants of the village to find out the suitable dates for the people to participate in the health camp. Once the dates were finalised the project field workers visited house to house to inform the people about the dates, the schedule and the purpose of the health camps were notified on a village notice board. Besides, the people themselves were involved in communicating about the health camp to the entire village. The village youth made the announcements in the villages on mikes. The mikes are usually kept in the village temples to convey messages, meetings and gatherings and to make other important announcements. A good rapport was established with the people as a survey had been conducted just prior to the health camps. The people were well acquainted with the survey team and the study. These efforts helped in greater participation of the children and the people in the health camps.

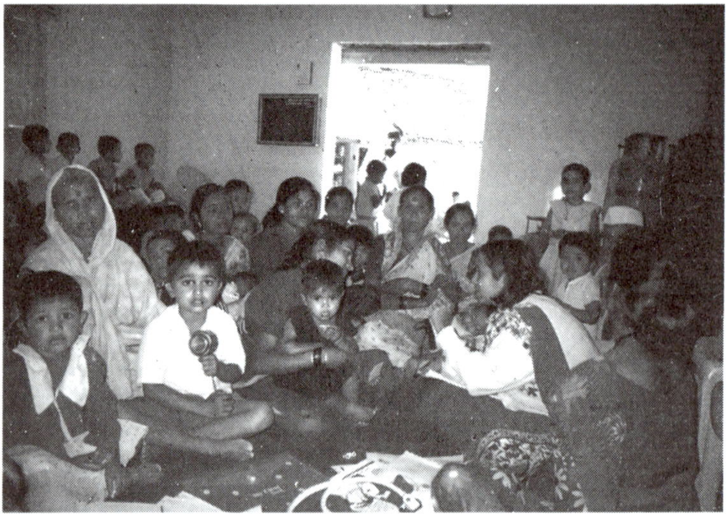

Plate 3: Health camp in village Mhaswandi

Each village is characteristically different. Certain villages have difficult geographical conditions. For example, Mhaswandi is a hill village with few hamlets right on hilltop while Kangar is a large village with dispersed hamlets. The average distance between hamlets being approximately six kilometres. While people from most of the hamlets brought their children right to the venue of the camp, attempts were made to cover all hamlets of a village with the help of organisation's office vehicle and by walking the distance where vehicle could not reach. The health camps were held in two to three different places in one village to make it convenient for the people from different and nearby hamlets to attend and so as to cover the entire village. The camps were held in places like office of women's committee, the *Samyukta Mahila Samiti* (SMS), the *anganwadi*, in school premises, courtyard of a house and the village temple.

Study Consent

An important part of the study was to obtain the consent of the people for the study. At village level verbal consent was obtained to conduct the health camps by holding community meetings prior to the activity. In addition, consent was also obtained from the parents before beginning the clinical examination. A consent form developed for the purpose was read out before the individual parents of the child participating in clinical examination. The consent was verbal with a written script.

The consent form stated the benefit as people getting to know the nutritional status of their children from a doctor, an opportunity to consult the doctor to know more about their child's health and to seek guidance for health problems. The form also stated that there was no specific risk associated with clinical examination except that it would consume time of the participants: the children and the parents or their care givers. That the clinical examination was based on observation of the child and interview of the mother or the caregiver accompanying the child in the camp and that no blood tests were involved. The consent form further stated that participation in clinical examination was voluntary and that the data and the identity of the participants would not be disclosed to anyone except to the research authorities that would make use of the data for generation of knowledge and for the purpose of community development. That the feedback of nutritional and health status of the child would be given to the parent or the care giver individually in the camp with a view point to generate awareness regarding their children's health status.

Data Analysis Method

As the data are quantitative, they have been analysed in SPSS (Statistical Package for the Social Sciences) version 14.

Analysis

The results of the study are shown in numbers and percentages presented below in Table 2.1. The table is divided into three columns. One column describes the clinical symptoms looked for during the examination. The results of which are divided into two: number and percentage of children having normal status and those with below normal or poor nutritional status as diagnosed by the physician.

The sample size differs for certain symptoms observed in specific age groups. The changed number is indicated in the table under the symptom.

Table 2.1

General clinical features and nutritional status of children surveyed[1]

(N=174 unless otherwise specified)

Symptoms		*Status*	
		Normal	*Poor*
General clinical features			
Appetite	N	Normal: 163	Poor: 11
	%	93.7	6.3
Thirst (in children above age 6 months) N = 144	N	Normal: 144	Poor: 0
	%	100.0	0.0
Dehydration N = 173	N	Absent: 167	Present: 6
	%	96.5	3.5
Urine	N	Normal: 169	Infection: 5
	%	97.1	2.9
Stool	N	Regular: 170	Constipation: 4
	%	97.7	2.3
Skin type	N	Normal: 131	Dryness: 43
	%	75.3	24.7
Activeness	N	Playing: 143	Weak: 31
	%	82.2	17.8
Response to stimuli	N	Present: 171	Absent: 3
	%	98.3	1.7

Cont'd...

...Cont'd

Other illness	N	Absent: 168	Present: 6
	%	96.6	3.4
Clinical features of PEM			
(a) Oedema, moon face	N	Absent: 169	Present: 5
(b) Hair changes to silky, sparse and get easily pulled out	%	97.1	2.9
(c) Growth retardation			
(a) Muscle wasting severe	N	Absent: 166	Present: 8
(b) Growth retardation	%	95.4	4.6
Diarrhoea (age 0–1 year) N = 36	N	Absent: 35	Present: 1
	%	97.2	2.8
Diarrhoea (age 1–5 years) N = 119	N	Absent: 119	Present: 0
	%	100.0	0.0
Clinical signs of nutritional deficiencies			
Eye sight (ophthalmic perspective) N = 169	N	Normal vision: 168	Poor vision: 1
	%	99.4	0.6
Vitamin-A deficiency	N	Absent: 163	Present: 11
	%	93.7	6.3
Vitamin-D deficiency N = 173	N	Absent: 165	Present: 8
	%	95.4	4.6
Anaemia	N	Absent: 143	Present: 30
(a) Low haemoglobin as observed from lower eyelid	%	82.8	17.2
(b) Skin features: Presence of pallor N = 173			
Ankle joint weakness	N	Absent: 161	Present: 13
	%	92.5	7.5
Skin infection	N	Absent: 168	Present: 6
	%	96.6	3.4
Worms	N	Absent: 149	Present: 25
	%	85.6	14.4
Dental caries (age 2 years onwards) N = 92	N	Absent: 92	Present: 0
	%	100.0	0.0

The common illnesses found among the children include cough, cold and fever. Illnesses that were found among one or two individual cases include heart problems, late milestones, cerebral palsy and speech problems.

The figures indicate that quite a large number of children are normal.

Table 2.2

Nutritional status of children aged 0–6 years as per anganwadi records pooled for all the villages

Total	*Category*	*Nutritional status*				
		Normal	*Malnutrition grades*			
			I	*II*	*III*	*IV*
908						
	Number of children	484	349	67	7	1
	Percentage	53.3	38.4	7.4	0.8	0.1

The ICDS concentrates on the age group 0–6 years and therefore the above data are for the same age group.

Child Mortality

There are seven cases of deaths (0.8%) during the period from January to December 2007 as recorded by *anganwadi* workers. The data were cross-checked with the people in the villages.

Discussion

The results obtained from clinical examination, the *anganwadi* records of nutritional status and child mortality indicate the following:

(i) A substantial number of children belong to normal nutritional status in the semi-arid rural area under study, as observed from clinical signs and symptoms.

(ii) The body parameters like appetite, thirst, urine, stool and normal body temperature are essential to determine the status of the health. The first indications about malnourishment in children are poor appetite, decreased activity and peevishness. An illness is usually associated. The observations based on these parameters indicate normal status of health of children under study.

(iii) Nutritional status of children based on weight-for-age as observed from the records of *anganwadi* indicates prevalence of malnutrition more in the first grade. With some more efforts and development inputs it is possible to bring down the level of malnutrition. The results from the *anganwadi* records and those from clinical examination show difference with regard to the percentage of normal children. This is largely because of the difference in the sample size of the two. The *anganwadi* records give a picture of the children of the entire village while the clinical examination has been conducted with children participating in the health camp. The records of *anganwadi* regarding grade II malnutrition (7.4%) match with the results of clinical features of PEM (7.5%) indicating that the PEM is comparatively less in higher grades. The two types of data (primary and secondary) help us in giving a direction for further development action.

(iv) Dryness as exhibited on skin is found in large number of children. Dehydration and diarrhoea, water-related problem, is almost negligible. Skin dryness, as observed in the present study, could be because of cold as part of the data collection took place in winter.

(v) Although anaemia can be stated with confirmation only with a blood test, the percentage of anaemia (17.2%) as observed from skin features and from lower eyelid is much lower in comparison with the NFHS data. NFHS-3 has classified anaemia into three categories: mild (20.8%), moderate (43.6%) and severe (2.4%) in rural Maharashtra, in the age group of 6 to 59 months (IIPS, Macro, Maharashtra report, 2008). The anaemia found in the present study corresponds to mild category of NFHS.

(vi) The prevalence of PEM is a cause of concern (7.5%) and needs urgent attention from health development perspective. PEM is a sum of many nutritional deficiencies specially protein and energy. As this is an important factor contributing to malnutrition it is suggested that studies based on anthropometric parameters and food intake be undertaken to look for and validate the results of PEM. There is a need to strengthen the component of nutrition education in government ICDS and where NGO initiated participatory growth monitoring programmes to combat PEM. According to NFHS-3, 42 per cent of children in rural Maharashtra are underweight (IIPS, Macro, Maharashtra Report, 2008).

(vii) The number of children having worms is comparatively low (14.4%). However, the current figures should be brought down with more efforts as worm infestation could worsen anaemia. Infestation with worms is often related to the non-availability of safe drinking water. More awareness needs to be provided regarding fetching water from safe source and the importance of water purification.

(viii) There is low prevalence of Vitamin-D and calcium deficiencies and skin infections like dermatitis. Dental health is good. No dental caries are observed.

There are very few studies available for comparisons that are undertaken among the pre-school children from the dry belt in the recent times. A study by Bhatia et al. (2007), conducted among the pre-school children from urban slums, has reported 37.4 per cent of the children having normal weight, a very high prevalence of anaemia (41.5%), prevalence of severe PEM observed in 7 per cent and worm infestation in 35.7 per cent. In comparison, the findings of the present study show a lower rate of malnutrition with more number of children having normal weight (53.3%), lower rate of clinical conditions like anaemia and worm infestation. Although there is difference in geographical area and culture of the community under study by Bhatia et al. (2007) and that of the present, poverty is a common factor in the two areas. The present study conducted among the rural population in the dry belt helps us to understand the effectiveness of water and soil conservation. The improved levels of child nutrition also indicate a declining trend in the levels of poverty.

Deaton and Dreze (2009) point out a sharp decline in the prevalence of clinical signs of nutritional deficiency, such as marasmus and oedema and a steady retreat of severe undernourishment in the last thirty years. The results of the present study point consistency with regard to these observations.

Water and Soil Conservation in Dry Lands

Programmes for water and soil conservation have been one of the major areas where long term development efforts of the government and the non-government organisations have been concentrated. Integrated watershed development has the potential to bring about a change in several areas affecting rural livelihood: irrigation and crop production, livestock extension, drinking water, sanitation, environment and social development contributing to enhanced child nutrition.

Role of Government

In the dry and semi-arid regions of the country, water and soil conservation has been initiated by government since independence. In Ahmednagar, a soil conservation scheme was initiated in the year 1958–59. Works like contour *bunding, nala bunding* and terracing were completed by 1970. Improvement in agriculture in the non-irrigated rainfed areas was a priority for sustaining rural livelihoods. Programmes with watershed development approach like Drought Prone Area Programme (DPAP) and Desert Development Programme (DDP) were taken up after independence to enhance agriculture productivity in rainfed and drought affected areas. Maharashtra, a state in the western part of India has a large drought-prone area (52%) and has faced recurrent droughts and famines (1907, 1911, 1918, 1920, 1972) (Samuel et al., 2007). After the drought in 1972 water and soil conservation work was accelerated in the semi-arid and dry land areas. Work on construction of village tanks, percolation tanks, soil bunds, check dams of different types was undertaken. Irrigation potential in the state and the semi-arid regions in particular was increased. In order to accrue more benefits, the government in 1983 adopted watershed development as a holistic concept and a programme of integrated watershed development was initiated. Under the programme it was decided to undertake soil and water conservation works from ridge to valley basis and according to the land use pattern. During the same time, the government also launched an aggressive programme of irrigation infrastructure. Development works in irrigation were taken up in all five-year plans. The first five-year plan (1951–1956) started work on the irrigation arena. Irrigation has assumed considerable significance at the state, regional and national levels. Surplus agriculture in the semi-arid and arid regions was possible as a result of all these efforts, like extensive irrigation and water and soil conservation measures including watershed development (Raju et al., 2004).

In order to further improve agriculture productivity in the region, watershed development programmes were introduced in these areas by Government of India. A programme on National Watershed Development Project for Rainfed Areas (NWDPRA) was launched on a massive scale during the seventh five-year plan period (1981–1986), covering different rainfed regions of the country (Narayanmoorthy and Kshirsagar, 2008).

Maharashtra is the first state to launch NWDPRA programme. Since its inception efforts have been taken by the government to make revisions and improvements in watershed development programme, based on the recommendations of commissions and committees (Irrigation Commission, Sukhtankar Committee, 1973, Dr Subramaniyan Committee, 1987, Prof. Hanumantrao Committee, 1994). The year 1992 marks the commencement of integrated watershed development programme in a revised manner (Government of Maharashtra, 1999).

Role of Non-Government Organisations

At about the same time when government programmes were implemented, a comprehensive programme, the Indo-German Watershed Development Programme (IGWDP), was conceived and initiated by Fr. Hermann Bacher, in Ahmednagar in the year 1989. The approach of IGWDP rested on community participation and self-help efforts organised around revitalisation of the local environment. IGWDP is a multi-sectoral programme that involves integration of different components of development as described earlier.

The IGWDP was implemented by the formation of an organisation – WOTR – established in 1993 in Ahmednagar to undertake watershed developmental activities in resource fragile rainfed areas. WOTR was started as capacity building agency for the IGWDP and is now a resource organisation in the field of watershed development.

People's Institutions

Formation of people's institutions is a part of integrated watershed development works. Under the IGWDP, as a part of the integrated and participatory watershed development, formation of people's institutions like, Village Watershed Committee (VWC), Women's Committees (SHGs and its apex body), the *Samyukta Mahila Samiti,* takes place as facilitated by the WOTR. These institutions take up the work of development activities through self-help in the village. People's institutions act as facilitating agents with greater participation of the people in decision-making and development processes through self-help. For example, in watershed project villages, it was observed that with the help of VWC, people take part in construction of water conservation structures and through collective efforts have brought dry agriculture land under irrigation. The maintenance of water conservation structures is also a joint responsibility shared by the village people through self-help. The work is facilitated by the VWC and is undertaken from VWC funds.

The roots of the self-help movement date back to 1990s. Promotion of savings habit among people, especially women, was initiated. It started with the concept of saving money in a group and utilising it by any one of the members for emergency purpose. The objective of the movement was to make women independent. Saving money was just one aspect but utilising the strength of a team (an SHG) for women's and community's own development was the real motto. Once the woman is independent by setting up a business and running own livelihood profitably she takes interest in participating in overall village development and her opinion is valued. Realising the strength of women, more and more banks are taking interest in linking up to lend the women self-help groups. National Agriculture Bank for Rural Development (NABARD) launched a two-phase programme: Micro Entrepreneurship Development Programme (MEDP) for SHGs, in 2006, which has completed two years (Jaipurkar, 2008).

Water Resources in Ahmednagar

The development efforts in the past fifty years led to increase in water resources in the dry district. A review of water resources in Ahmednagar indicates that about 32 per cent of the cultivated area is under irrigation, out of which 71 per cent is under well irrigation (including lift irrigation) and remaining area is under canal irrigation. In the northern part of the district there are three major irrigation projects, namely, upper Godavari, Pravara and Mula having 2, 18 and 28 per cent of the irrigated areas respectively. In the south-western part of the district there are two command projects (namely, Ghod and Kukadi), which are having 15 per cent of the irrigated area. Besides this, there are seven medium projects and 90 minor projects covering 16 and 5 per cent of the total irrigated areas respectively (http://www.kvk.pravara.com/ distprofile.htm).

Status of Water and Soil Conservation in Study Villages

In four out of five study villages, viz., Mhaswandi, Gunjalwadi, Kangar and Karjule Pathar, integrated watershed development work has been undertaken based on IGWDP by local organisations with capacity building and technical support provided by WOTR. The work was completed during the years from 2000–02.

The total land area treated under watershed development in four watershed villages is 5,497 hectares. The figure exceeds the total village land area of the four watershed villages. This causes an impact on the adjacent villages, which too get the benefit of watershed development. The total population of the watershed villages is 6,809 that have benefited from watershed development. Watershed development has led to the economic growth in these villages with reduction in water scarcity for drinking and agriculture. In Kalas Pimpri integrated watershed development work has not been undertaken. However, minor water and soil conservation work has been facilitated by the government departments of Agriculture and Water Resources, Social Forestry and by Village Panchayat. The total population of this village is 1,187. There are 20 soil bunds and six cement dams in the village. Continuous Contour Trenching (CCT) work has been carried out on 10 hectares of land. Tree plantation work has been carried out with more than 10,000 trees planted in this village. These activities are carried out in an integrated manner in integrated watershed development. Watershed development entails capturing the rainwater where and when it falls. With the help of contour trenching on hilltops and slopes, rainwater is made to percolate into the soil rather than flow downhill and be lost forever. Tree plantations, which have been done in lakhs in every watershed village, help to prevent soil erosion. Check dams, and other methods of water and soil conservation, impede the flow of rainwater. All this increases the water table level and stops the wells from going dry. The wells in the villages, where integrated watershed development has been implemented contain water even in the summer, which was not the situation prior to watershed development. There is a presence of base flow in the streams of the villages even in summer, as a result of watershed measures.

Safe Drinking Water and Sanitation

Positive health and well-being can be achieved only with safe water. Water that is easily available and accessible, adequate in quantity and free from contamination is considered as safe. National Water Supply and Sanitation Programme was initiated by Government of India in 1954, with the object of providing safe water supply and adequate drainage facilities for the entire urban and rural population of the country (Park, 2005). The improvements in water resources in turn have contributed to the success of the programme.

Plate 4: Year round availability of water in watershed villages

Plate 5: Crop production in a watershed village

In all the study villages piped drinking water is available. Village level observations indicate that about half of the population in all the study villages have easy access to drinking water within the household premises. The efforts to bring the water pipelines in the villages have been undertaken by the villagers. The process was initiated by the VWC after the implementation of watershed development in their villages. Reduction in time, energy and the distance travelled in fetching water by the women was a major resulting change. The time and energy thus saved have consequently helped in contributing towards better child care and development.

Chlorination of drinking water is done on a regular basis. The village panchayat and the members of the village watershed committee undertake the responsibility of chlorination of water in the wells. The role of the government and the non-government organisations has been important in generating awareness regarding practices of safe drinking water. There is greater awareness among the people in making water safe for drinking purposes. The village people were found to be particular about chlorination of water. A bottle of Medichlor was found in many houses and chlorination of drinking water takes place at household level on daily basis. During water scarcity, the village panchayat makes the provision for the availability of water to the entire village by means of water tanker. Filtration of tanker water and its chlorination at household level is the usual practice followed by most of the people, as the tanker water is turbid.

Home toilets have been constructed in all four watershed villages. The people in watershed villages were found to be taking keen interest in curbing open defecation. The Zilla Parishad (ZP) and NGOs involved in watershed development have spread awareness regarding the effects of open defecation. The government is aiming to make rural areas free from open defecation. The practice of open defecation has been reduced as observed in the watershed villages. The people make use of latrines. The latrines are constructed outside the house or at some distance from the house. Watershed villages have visitors coming to see the impacts of watershed development. There are students, members of donor agencies, researchers, people from government and non-government development organisations and other visitors from different parts of the country and foreign countries. 'Utilisation of toilets' for defecation is looked upon as a status symbol. The VWCs in watershed villages have levied fine to curb open defecation. They were keen about obtaining award for being an ideal village *(Adarsh Gaon Yojana)*. As a result open defecation has come down drastically, a step contributing towards women and child health. In the country, Maharashtra tops the charts in rural sanitation. The state now has a number of open defecation-free villages, called *nirmal grams* (clean villages) and it was also observed in the study villages. In 2005, 13 PRIs in Maharashtra could achieve total sanitation. The number rose to 1,912 in 2007 (www.financialexpress.com).

Cropping Pattern in the Study Villages

Ahmednagar has a highest geographical area of 17 lakh hectares. It possesses fertile soils of different types. A total of 30 per cent of the total cultivable area is under irrigation. A system of multiple cropping is prevalent. Multiple cropping yields different types of produce, resulting in a balanced diet for the family (http://www.kvk.pravara.com/distprofile.htm).

In the villages, bajra is the staple crop. Onion, potato, tomato, groundnut and cotton are the cash crops grown in large amount. It is an observation that watershed development has brought about an increase in the cropping pattern. Among all cash crops, onion and cotton are the most important commercial crops grown more widely. They have yielded profits in crores. Cotton is a dry land crop but onion production is dependent on water. Watershed development has led to increased production of onion. The villages have become prosperous due to higher yields of onion. Onion leaves *(Kanda pat)* and bajra are rich in iron. Pulses like *Tur* (Red gram *dal*/Cajanus cajan), *Harbhara* (Bengal gram *dal*/Cicer arietinum), *Urad* (Black gram *dal*/Phaseolus mungo Roxb), *Mug* (Green gram *dal*/Phaseolus aureus Roxb) *and Matki* (Moth beans/Phaseolus aconitifolius Jacq) and vegetables like fenugreek (*Methi*/Trigonella foenum-graecum) are grown usually and widely. There is a practice of using soybean oil in diet. Soybean oil is purchased from shops. The crops are rich sources of iron and protein (Gopalan et al., 2007). Other crops taken in small amount are jowar (Sorgham vulgare), wheat

(Triticum aestivum), paddy (Oryza sativa) and corn (Zea mays). Bajra bread and vegetable item made up of onion leaves, and rice and *dal* made up of red gram, form the staple food items.

Dairy Development

Watershed development largely contributed to livestock development. Increased agricultural productivity leading to fodder development and year-round availability of water for livestock made purchase of cross bred cows by farmers leading to increase in milk production and more number of dairy co-operatives. The dairy co-operatives have developed as a secondary movement benefiting the farmers. There are 1,281 co-operative dairy societies in Ahmednagar, providing impetus to the rural economy (http://www.kvk.pravara.com/distprofile.htm).

In all, there are five milk dairies in the study villages with a collection of approximately 17,000 litres per day.

Women in the Development Process

Women institutions are functioning in watershed villages over a period of five years.

In the study villages, there are a total of 41 women self-help groups (SHGs) functioning, out of which 39 are from watershed villages alone. There are two women committees, the *Samyukta Mahila Samiti* (SMS), in the watershed villages. The women members of SHGs and SMS take part in decision-making and participate in village development processes. One of the most recent examples is observed during the village meeting conducted during the fieldwork for the present research when women from a watershed village were found well represented in the meeting, expressing their opinions, asking queries with confidence and extending support to the study.

Every village has a single Gram Panchayat. The villages, Karjule Pathar and Gunjalwadi, are two separate revenue villages, having two separate *anganwadis* but fall under one Gram Panchayat. Out of the total 32 Gram Panchayat members in these villages, 11 (34.4%) are women. The women, particularly in watershed villages, are more organised due to their institutions, the SHGs and the larger groups, the SMS take active part in village development processes. The SHG women have closed down liquor shops and put an end to private moneylending business *(savkar dhanda)*. This also indicates that women are not just involved in putting across their opinion, rather they have brought about the change themselves. These aspects that have contributed to women development have in turn influenced upon child development.

The SHG women take part in preparation of *anganwadi* food for pre-school children. The kind of food to be cooked is decided jointly by the women members and the responsibility is shared in rotation. The *anganwadi* food includes items like *mung khichdi* (rice cooked with green gram dal), *harbhara usal* (legume preparations), and *lapsi/sheera* (a sweet item made up of grounded wheat fried in ghee, that is saturated fat mixed with jaggery).

The involvement of women in child nutrition is again a step towards child development. Today, the women organisations are involved in child development through raising awareness regarding child nutrition.

Development

Integrated Child Development Services (ICDS) is one of the important schemes for pre-school children's health. It was launched in the year 1975 all over the country. It is one of the world's largest government programmes for early childhood development. The services are provided through *anganwadis* in the village. A package of six services is provided under the ICDS: supplementary nutrition, non-formal pre-school education, immunisation, health check-up, referral services, nutrition and health education.

Supplementary nutrition includes supplementary feeding and growth monitoring, and prophylaxis against Vitamin-A deficiency and control of nutritional anaemia. The children below the age of six are identified. The supplementary feeding support is provided for 300 days in a year. With this the *anganwadis* attempt to bridge the protein energy gap between the recommended dietary allowance and average dietary intake of children. The programme aims to provide the right supplementary nutrition to children in the age group of six months to six years and is the world's biggest scheme, covering 58 million children.

Growth monitoring and nutrition surveillance are the two important activities that are undertaken. Children below the age of three years are weighed once a month and children between three and six years are weighed every quarter. Weight-for-age growth cards are maintained for all children below six years. This helps to detect growth faltering and assessing nutritional status. Besides, severely malnourished children are given special supplementary feeding and referred to health sub-centres and primary health centres, as and when required (www.icds.nic.in).

Integration of Water, Development and Child Nutrition

An integration of water, development and child nutrition is also observed during the fieldwork. For example, the women members of the SHGs in the villages are found involved in cooking the *anganwadi* food for children. As these women are also the mothers of children aged 0–5 years, they take better care and see that the children receive a balanced diet. In the watershed villages the SHGs are initiated by WOTR.

Village Level Development (Infrastructure)

A strong village infrastructure with services available within reach contributes in a big way to overall development of the village impacting upon child nutrition.

Health Care Services

All the study villages have government primary health sub-centre located in and around the village, itself. For all villages, except Mhaswandi, basic health care services are available within a range of 2 km. Mhaswandi is a comparatively remote village. There, the health sub-centre is about 5 km away. Two villages, Kangar and Kalas Pimpri, have local private doctors attending more than three times in a week. The government primary health centre (PHC) is at a distance of about 16 km for all villages. There are private clinics closely connected within a range of 10 km. Also, there are traditional healing practitioners of different types in the villages, like herbal specialists, bone setters and magico-religious healers in the villages.

All the villages have *anganwadis* – the child care centres. Kangar, being a large village and having dispersed settlements, has four *anganwadis* while the other villages have one each. There are thus a total of eight *anganwadi* centres in all the villages.

Transport and Communication Services

The location of three villages – Mhaswandi, Karjule Pathar and Gunjalwadi – is close to Pune-Nasik highway. Pune and Nasik are two large cities and district headquarters in Maharashtra. Although not very close, these villages are well connected to these city places by means of transport, both public and private. Motor bikes and bicycles are commonly found. In Mhaswandi, which is a little remote, people help each other and plan out work in outside places with the people who have bikes. Besides, private trucks and tempos coming into the villages and carrying food grains are also of great help to move out to city and district places. All the villages have either public and/or private telephone facility. Some people, like the ICDS workers, village development workers, school teachers, local leaders and the large farmers have mobile phones which help them to connect to the outside world and comparatively better developed places and avail the necessary benefits. Postal facility is available within a range of 5 km. The average distance of the study villages from the *taluka* (sub-district) place is 27 km and it is 80 km from the district headquarters.

Education

Education is one of the most important factors affecting child nutrition and overall development. It was observed that most of the mothers had completed schooling up to the seventh standard. This has contributed to better awareness regarding immunisation. It is observed that more than 80 per cent of the children have been covered under the immunisation programme. Education of the mother has a vital role in child nutrition and in bringing down the mortality rate (Deshprabhu, 2001). Availability and accessibility to education speaks of the village development. In all the study villages, primary school (1–4 standards) facility is available. Three villages, viz., Mhaswandi, Kangar and Gunjalwadi, have secondary school (1–7 standards) facility. The children from villages – Kalas Pimpri and Karjule Pathar – have to travel up to two and a half kilometres to get down to the nearest secondary school. Village Kangar has higher secondary (1–10 standard) school facility available while for all other villages it is available at a distance of 3 km.

Public Distribution System (PDS)

Every village has a ration shop. A PDS in place is indicative of village food security. People can avail of the food grains at low cost when in need in the village itself. This has also contributed in bringing down household food shortage in the villages.

Conclusion

The study and its results have added to the database for the organisation that initiated watershed development in the villages studied from the point of view of child nutrition as no such impact assessment study was carried out earlier in the region. The study indicates that malnutrition is existent but the situation is not so bad. With more concerted efforts the desired change in child nutrition can be brought about. It is suggested that watershed development programmes be strengthened with component for child nutrition. Anaemia should not be ignored and efforts to bring down the rate are essential. It is suggested that an estimation of haemoglobin should be done to assess the degree of anaemia.

'Water development' is truly a leading factor towards human development. Water contributes to livelihood. Water and soil conservation and particularly the integrated watershed development have brought about agriculture and livestock development in semi-arid and dry land regions. Water has been the means around which women development has taken place through their organisations. A study based on four villages is no doubt a small-scale study but gives some pointers for understanding and undertaking future actions for intervention. Child nutrition is a measure of human development. No programme or a factor works in isolation towards development. Water is a base of most of the factors interplaying in child nutrition. The factors working at macro-level impact on development that takes place at micro-level. The efforts and actions at country level or regional level influence upon the actions at village level. To list out a few as observed from the study: the contributions of water, allied agriculture and dairy development, development schemes in operation and inputs, the efforts of government and non-government organisations into it, people's awareness and practices of sanitation, people's initiatives in village development, availability and accessibility to development services (village infrastructure), connectivity of the villages to the outside world, the lead role of women in development processes and women's education are some of the factors that have laid a positive influence on the nutritional status of children. Adequate availability of water for sanitation as a result of water and soil conservation measures, government assistance for sanitation under the rural development schemes, easy availability of cash as a result of economic growth in watershed villages and people's interest, initiatives and participation in making their village an ideal are some of the factors that have contributed to safe sanitary practices and consequently to child nutrition and health.

More child nutrition inputs to bring down the problems like anaemia, worm infestation and protein energy malnutrition are essential for further improvements in the area.

Watershed development programme should not be a mere technical intervention but must incorporate human development aspects like child nutrition right from the planning stage. This is because malnutrition is a foremost problem in India and also an indicator of human development. Water development programmes strengthened with efforts directed towards child nutrition are likely to make a headway towards child development. An active participation of watershed committees and women committees in the process of growth monitoring and in setting the strategies to improve upon the nutritional status of children should be encouraged. This way an integration of watershed development and child development programmes like that of ICDS could be achieved to work towards a common goal of reducing malnutrition and contributing to child development.

Note

1. Reference for listing clinical symptoms: Park K. *Preventive and Social Medicine*, Banarasidas Bhanot, Jabalpur, 18th edition, 2005.

3

Investigational Intricacies

As evidenced through records of WOTR, integrated and participatory watershed development has greatly contributed to agricultural production by means of increase in water level of the wells. Livestock development was achieved by increase in fodder produce by plantation of grass varieties on common lands and purchase of cross-bred cows. The milk production was consequently increased.

The unpublished reports of Sampada Trust for the periods April-October 2006 indicate that because of the participatory method of growth monitoring, awareness has been generated among the mothers and the *anganwadi* workers regarding the nutritional status of children aged 0–5 years.

Based on the records of WOTR and Sampada Trust, the following hypothesis was developed for testing in the field: "Integrated watershed development, strengthened with participatory child growth inputs, leads to reduction in the number of malnourished children aged 0–5 years in the rural semi-arid region".

The objectives of the study are as follows:

1. To examine the nutritional status of children aged 0–5 years based on weight and height-for-age.
2. To understand the feeding and dietary patterns of the children.
3. To understand the watershed-related and social/cultural factors influencing the nutritional status of children.
4. To understand the social and economic status of the families and the profile of mothers of the children.

The basis for testing the hypothesis was field experience, which indicated increase in land productivity and fodder production, thereby increase in crop production and milch animals. The increased food and milk availability and the inputs of participatory growth monitoring of children aged 0–5 years are likely to raise and improve the weight and height of these children as shown in the Figure 3.1 below:

Figure 3.1

Watershed development and participatory growth monitoring leading to improved nutritional status of children aged 0–5 years

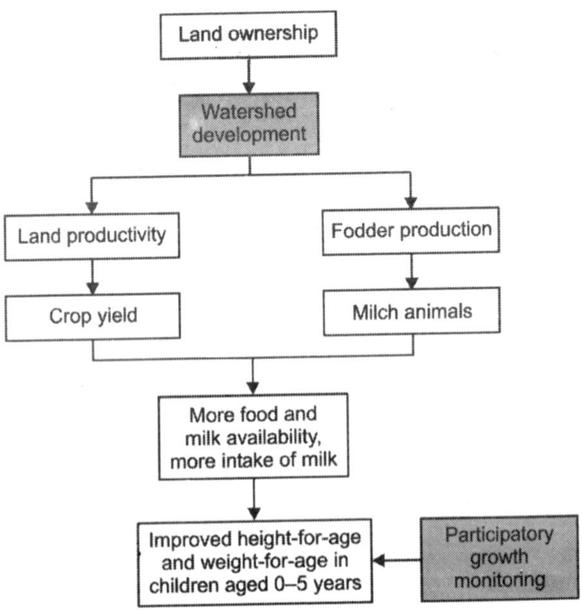

Figure 3.2
Factors influencing the nutritional status of children

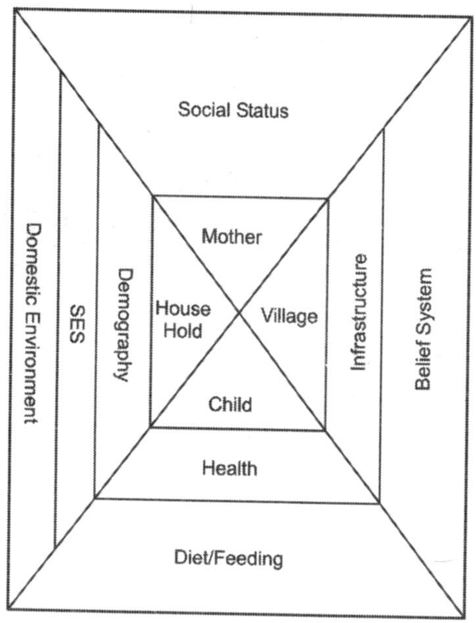

In order to test the hypothesis, the study was conducted using three comparative groups.

The purpose of the study addresses two important research questions:

1. Whether availability of more food and milk for family has influenced the enhancement of nutritional status of the children in the age group 0–5 years? If yes, how?
2. Whether increased awareness of the nutritional status of the children brought about through the work of growth monitoring has influenced the enhancement of nutritional status of the children in the age group 0–5 years? If yes, how?

Lack of Background Literature

There is a dearth of studies measuring the impacts of watershed development using anthropometric indices against a holistic understanding of village, household, women and child development profile, carried out among the children aged 0–5 years in the rainfed, arid and semi-arid region of rural western Maharashtra in a comparative perspective. The studies assessing the importance of growth monitoring as a component of watershed development are still rare. There are also very few nutritional studies based on anthropometric and dietary assessment of children aged 0–5 years in the rainfed rural-arid and semi-arid regions of western Maharashtra in the past five years in particular and even many years beyond. As impact assessment of watershed development in terms of child nutrition has not been much taken up, there is a lack of baseline data gathered from this perspective in the rainfed rural-arid and semi-arid western Maharashtra. Hence, there was no data available that could help as previous experience to build up the present study and no data available to compare the results of the present study.

As discussed, child nutrition is governed by many factors which also contribute to community development. The present study focuses on understanding the status of development with parameters that intermingle with watershed development and child nutrition, that is, they are affected by watershed development and are contributing factors of child nutrition. The parameters as used in the present study are related to village level and household level characteristics of children aged 0–5 years under study, characteristics of women as mothers of these children, and those of individual children.

A study was undertaken to test the hypothesis, understanding the impacts of integrated watershed development and NGO-initiated participatory growth monitoring on the nutritional status of children aged 0–5 years. The study is comparative in nature. Due to the lack of baseline data, a control group was used for comparison.

The study team composed of the principal and an associate researcher with anthropology background, an assistant for monitoring the survey with social work background, four field investigators who remained throughout the study, and the students of Rural Chroniclers, WOTR (year 2007–08). Rural

Chroniclers was a one-year training course conducted by WOTR for the educated and unemployed youth of the villages where watershed development is implemented. These students, after completing the course, helped the organisation to document the village development processes at the village level itself. Statistical consultancy was sought for sampling, development of interview schedules in the initial stages, data coding and analysis. The research advisors having expertise in watershed development, anthropology, and food and nutrition have always been there for guidance and all support required throughout the study.

Review Meetings

Review meetings have been an important element of the study. The discussions and review by experts, senior staff of WOTR and peer researchers for building up the foundations of the study have taken place through review meetings. The concept note, the research proposal, the tools and techniques, and the consent forms developed were critically reviewed before entering into the field for actual data collection.

Training of Researchers

Anthropometry and diet survey methods have been used in assessing the nutritional status and food intake of children under study. The researchers, who were going to be the trainers of field investigators, organised two days' training programmes to get themselves refreshed and updated in the two important methodologies. The research advisors with their expertise in the respective fields conducted the training in theoretical aspects of anthropometry and diet survey and gave practical demonstration.

The training programme on anthropometry was conducted by our research advisor, as expert in the subject. The staff of WOTR having children aged 0–5 years was requested to participate in training programme on anthropometry. The researchers had a learning experience of taking body measurements of children. The difficulties, the responses of children to their body measurements, dealing with such a small age group of children, and the time that would be required for anthropometry was judged on the basis of this practice session held during the training programme. At the same time, experience was gained in taking accurate body measurements and recording. Various steps and precautions were discussed while taking the measurements of small children.

The training programme on diet survey (24 hour diet recall method and food frequency) was conducted by other research advisor, also expert in the subject. We practised taking diet survey of our colleagues at our research unit, WOTR, Pune and tried to anticipate the difficulties that would crop in.

Before proceeding to the methodological discussion, it would be important to mention that any reference to watershed development related to the present study made in the text refers to integrated watershed development based on IGWDP approach and a reference to growth monitoring refers to NGO-initiated participatory growth monitoring. The components of integrated watershed development based on IGWDP and those on participatory growth monitoring have been spelt out earlier.

Comparative Study Groups

The comparison is carried out at three levels in the study. There are three different population groups. The two of which are project population groups. The third one is a control group.

Group 1: A group of villages where integrated watershed development has been implemented and participatory growth monitoring activity undertaken (WSD+GM).

Group 2: A group of villages where integrated watershed development has been implemented but no growth monitoring activity undertaken (WSD).

Group 3: A group of villages where neither integrated watershed development nor NGO-initiated participatory growth monitoring activity is undertaken (No Input).

Research Site

The study is conducted in twelve villages of Ahmednagar district, Maharashtra, India. The district is a rainfed, drought-prone, semi-arid region. The rainfall is uncertain. The average annual rainfall in the district is 501.8 mm. Tables 3.1 and 3.2 give information on the villages under study and the talukas to which they belong.

Table 3.1
Study villages

Comparative group	*Villages*	*Taluka*	*Watershed implementing organisation*	*Year of initiation and completion of watershed development*	*Year of initiation of growth monitoring*
WSD+GM	Bhoyre Pathar	Ahmednagar	Don Bosco Bigar Sheti Sahakari Patsanstha Maryadit, Ahmednagar	1990–1994	June, 2006
	Gunjalwadi	Sangamner	Social Centre	1996–2002	
	Mhaswandi	Sangamner	Sangamner Bhag Sahakari Sakhar Karkhana (SBSSK), Sangamner	1993–2000	
	Vaiju Babhulgaon	Pathardi	WOTR	1996–2002	
WSD	Kangar	Rahuri	Rahuri Sugar Factory, Rahuri	1994–2001	Not applicable (NA)
	Karjule Pathar	Sangamner	Social Centre, Ahmednagar	1996–2002	
	Sheri Koldara	Parner	Social Centre, Ahmednagar	1994–2002	
	Tondoli	Pathardi	Medical Mission	1997–2001	

Cont'd...

...Cont'd

No Input	Astagaon	Parner	Not applicable (NA)	Not applicable (NA)	Not applicable (NA)
	Jachakwadi	Akole	NA	NA	NA
	Kalas Pimpri	Pathardi	NA	NA	NA
	Walwane	Parner	NA	NA	NA

Table 3.2

Average rainfall in study talukas

Name of taluka	*Average rainfall (mm)*
Akole	508.9
Ahmednagar	531.1
Parner	486.9
Pathardi	580.4
Rahuri	455.0
Sangamner	416.6

(http://www.ahmednagar.nic.in)

Selection of Study Villages

For the present study, the villages which have completed at least five years with watershed development, and the watershed villages where growth monitoring has been initiated, were selected. Experience suggests that after successful completion of any watershed development activity, it takes at least five years to harvest its complete potential benefits. A representation of different talukas from Ahmednagar and watershed implementing organisations was ensured to have a wider coverage of the district encompassing behavioural diversity in the data.

A major methodological hurdle was experienced in the selection of 'No Input' group of villages with respect to watershed development. A survey of twenty villages was carried out to select the non-watershed villages. Initially, the 'No Input' group was defined as the one with the villages where

Figure 3.3

Map of India highlighting Maharashtra state

Figure 3.4

Map of Maharashtra state highlighting Ahmednagar district

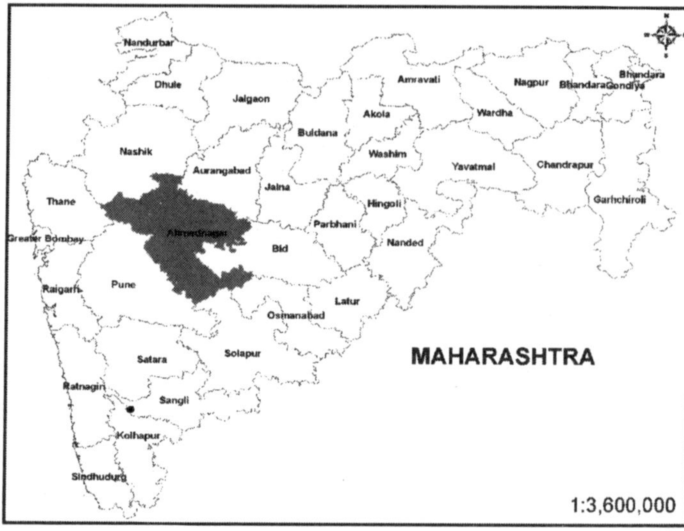

neither watershed development nor any water-related work like conservation and harvesting has been undertaken. However, no such village was found after moving in twenty villages from different talukas in Ahmednagar that could fulfil the selection criteria. In all the villages surveyed for the selection, water and soil conservation work initiated by government has been carried out. The 'No Input' group therefore included those villages where integrated watershed development based on IGWDP has not been implemented. There was no problem with regard to the exclusion of NGO-initiated participatory growth monitoring in the 'No Input' category.

Water and soil conservation structures like continuous contour trenching (CCT), small mud dams *(mati bandh)*, small cement dams (*cement bandhare*), percolation tank *(pazar talav)*, check dams, other minor dams *(vasant bandhara)* and farm bunds *(shet tali)* have been built in the 'No Input' villages. These structures have been built through Agriculture Department, Forest Department and Zila Parishad. In some of these villages, tree plantations have also been carried out.

Figure 3.5

Map of Ahmednagar district highlighting study villages

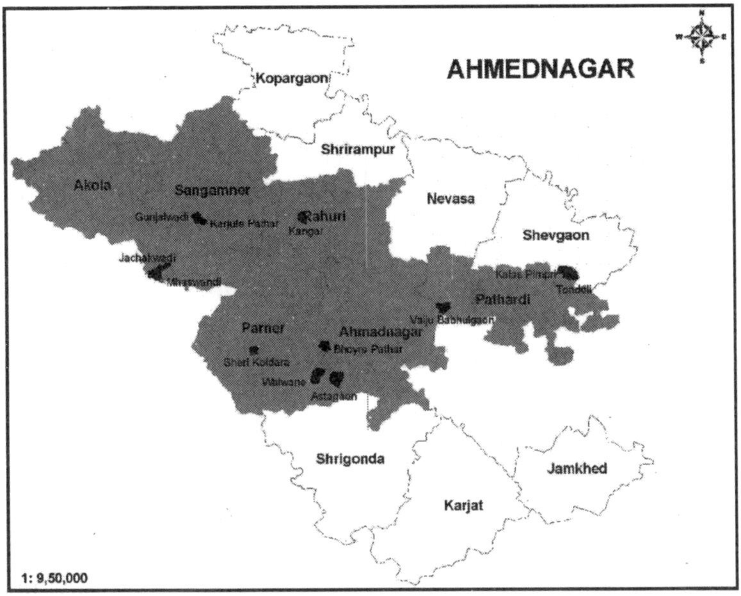

Sampling

It was decided to achieve a total sample size of 300 children who have completed 0–5 years (up to 71 months) of age. In order to achieve the above target, considering the dropout rate and migration, a list of additional 175 children was provided. The distribution of the covered sample size in the three comparative groups is as follows:

Table 3.3

Sample size of children in the study villages

Comparative group	*Total number of children in the villages*	*Sample size*	*Sample size covered*
WSD+GM	321	142	111
WSD	733	190	127
No Input	528	143	116

For selection of the sample, all the children aged 0–5 years in the village was considered as the universe. The *anganwadi* caters to the children of age 0–6 years. A complete list of children in this age group was obtained from the *anganwadi* centres.

From the *anganwadi* list, children completing 0–5 years were sorted out. For the purpose of sorting, as a rule, all children who were above five years and six months were discarded from the list. Children who had completed six years of age were not included in the sorted list. The sorted list then formed the sampling frame for selecting the required sample size for the study.

The sample size has been derived by using simple random sampling method.

Tools and Techniques of Data Collection

The study employs quantitative research method with the purpose of testing the hypothesis. The quantitative data have been collected using structured interview schedules. Five different types of schedules were used for data collection as follows:

1. Village Profile
2. Household Schedule
3. Schedule for Children
4. Mother's Profile
5. Anthropometric Schedule

Through the village profile schedule the information that was gathered was related to the infrastructure and services available and accessible to the people whose children have been covered under study. These include the key informants of the village such as *gramsevak*, *talathi*, *anganwadi* workers, school teachers, *sarpanch*, members of SHG, president of village watershed development committees (VWC), youth group leaders, police patil, etc.

The household schedule was aimed to collect data related to household socio-economic characteristics of children under study by interviewing the head of the household in most cases, the parent of the child, who may or may not be a head of the household or any other elderly member of the family in cases when the head of the household was not available.

The schedule for children focused on collecting information regarding child care and feeding practices, health, immunisation, dietary assessment based on 24 hours diet recall method and food frequency. The data were collected by interviewing the mothers or the care givers like a grandmother of the children under study.

The mother's profile schedule gathered information related to education, occupation, age at marriage, social status and health characteristics from the mothers of children under study.

The anthropometric schedule was used to record the anthropometric (body) measurements of the child, viz., weight, height, head circumference, chest circumference and mid-upper arm circumference.

Study Consent Form

A consent form was designed to take verbal consent of the study participants for taking part in the interview and to request and take consent of the parent/care giver for participation of children in anthropometry camp.

The consent form briefed the participants about the details of the study, the purpose, the methods and the procedure that would be followed in the study, time required for the interview and anthropometry camp, maintenance of confidentiality of the data and the identity of the participants, voluntary participation in the interview and anthropometry camp, risks occurring from participation in the study and the benefits.

Contact Card

A contact card was developed for the people to contact the office and the researchers whenever they wanted to clarify doubts or get more information on the study, if necessary. The card contained name, WOTR office contact information, and name and other contact information of the researchers of the study.

Selection of Field Investigators

A certified course was conducted by WOTR to strengthen development of the rural youth. The course named as 'Rural Chroniclers' was sponsored by the University of Pune, Pune. It provided one-year training in documentation of village development processes. The young girls and boys from the villages, who had completed basic schooling and were interested in reading and writing, were given opportunity to undergo the training. The rural youth from watershed villages were given priority. The syllabus of the course was application-oriented and equipped the students to acquire employment where documentation skills were required.

The present study involved village people in data collection. One was a group of young boys and girls from watershed villages, the students of Rural Chroniclers and WOTR. They had completed their schooling, up to 10th standard. The other was an elderly woman health volunteer *(Mahila Pravartak)*, involved in undertaking growth monitoring of children aged 0–5 years. She has undergone several training programmes conducted by WOTR and Sampada Trust in varied subjects. She was assigned the task of data collection in her own village, where she could be comfortable. She is a member of SMS and is highly respected in the village.

A few students of Rural Chroniclers course were nominated by the course coordinator, WOTR, who had an understanding of the students' capabilities and interest in undertaking tasks like that of data collection for a study.

The purpose of selecting village people for data collection was manifold. One was that the data collection for a study would help strengthen their documentation capacities further. The others were that the trained village youth and the village woman would get an employment opportunity. Besides, it would generate awareness of subject matters like nutrition and development among the people who are a part of arid and semi-arid, poverty-stricken and drought-prone villages.

Designing of Interview Schedules

The interview schedules were designed after conducting the ground work over a period of four months. The ground work involved a review of literature, discussions with different groups of people, viz., the research advisors, senior staff of WOTR, colleagues from research unit, WOTR, Pune office, and field visits so as to have a sense of the community and the villages. The interview schedules developed by WOTR for baseline studies and those by National Family Health Survey-3 (NFHS-3) were reviewed and were of great help. They offered insights for capturing watershed factors and child health aspects respectively.

An important element of consideration in the designing of interview schedules was that they were designed in a manner that would be simple for the village people to ask questions and record the responses rather than merely from the point of view of statistical data analysis.

A pre-testing of the schedules developed was carried out in the field. It was done in village Damalwadi, Pathardi, *taluka,* Ahmednagar. Damalwadi is about 40 km away from Ahmednagar. The experience of pre-testing helped in contextualising and making the schedules more focused. The initial drafts of the schedules were also revised. The interview schedules were discussed with the field investigators during the training programme. They were again field-tested by the investigators as a part of field practice. The experiences gained were incorporated in finalising the schedules.

The interview schedules were initially prepared in English language. They were then translated in Marathi, the local language in the villages.

Four different types of interview schedules were developed for study as follows:

(A) Household Schedule

(B) Village Profile

(C) Mother's Profile (Profile of Women as Mothers of Children 0–5 Years)

(D) Children's Schedule

Household Schedule

Some of the parameters used to develop a family/household profile of the children and their families are as follows:

(i) Infrastructure
(ii) Dependants in a household
(iii) Basic livelihood needs: housing, drinking water, fuel
(iv) Household assets
(v) Production and income-generating activities
(vi) Livelihood/food security: migration, food shortage, loans

Infrastructure

(i) *Roads, Transport and Communication:* Remote and isolated areas are often the areas which have more number of poor people and lack development. Means of transport and communication are basic to economic and social development of a region. Physical access to roads and mass communication has an important bearing on livelihood activities in rural areas. Access to health facilities and other basic needs is critical for all. A strong infrastructure of roads, transport and communication is one way of reaching out to poor. The poor can access basic services based on their needs. Transport and communication are inextricably linked to development. They open up opportunities for production and marketing (IFAD, 1992).

(ii) *Basic Services: Health Care and Education:* Availability of basic services like education and health care reflects development of a population. Besides, distance travelled by the people to seek these services also gives a fair picture of the development of a particular area.

Unavailability of primary health care and hospitals poses barriers to access health care services. Emergency services, for example, for delivery and other care, can help the rural people as they have to travel long distances to *taluka* level and district places. Besides, satisfaction in the quantity and quality of health care services is also an important factor which causes a barrier in the access of these services.

Education is a very important factor in the process of development. Educational status of the rural people itself speaks of the level of development. Education is an indicator of development. The impacts of women education on health need no explanation. Women with more schooling have found to be with less decline in fertility levels and change in reproductive behaviour.

(iii) *Other Services: Markets, Shops:* Weekly markets play a very crucial role in the rural life. The linkages of the villages to market places and the availability of shops indicates the level of development of a region. In rural areas, weekly markets have remained important from the point of view of trade since long. In the absence of retail shops in the villages, the villagers make the necessary purchases from the nearby weekly markets. The local economy revolves around the market. A number of micro-enterprise activities through SHGs and other livelihood groups take place in the market. Development in infrastructure like roads, transport and communication has also increased the scope of market. Markets are the rural economic growth centres. Watershed development has brought about a new prosperity in the lives of the rural people. Similarly, infrastructure development has improved the buying and consumption capacities of the rural people.

The availability of wide and different types of shops in the village reflects the changing demands of the people. Local shops play an important role. They supply goods to the people, contribute to the quality of life, provide a social focal point, encourage the population to stay and are a pre-requisite for the development of other diversification activities.

Bank and credit system has also had an impact on economic growth of the rural population. Availability of credit contributes to women empowerment. Availability of bank loans, as against the non-institutional mechanism like the moneylenders, significantly contributes to the economic growth of the rural poor. The poor can avail of the government subsidies through bank loans on certain productive activities. Institutional credit also contributes to poverty alleviation.

(iv) *Animal Care Services:* Livestock raising has a very important role to play in the life of the rural poor. Not only it helps in earning a livelihood, but also serves religious purposes in one's rural life. Many of the rural poor are engaged in livestock husbandry. Since disease is an important constraint on livestock productivity, development projects that improve veterinary services, or make them more available to the target group, can have a significant impact on rural poverty alleviation.

Dependents in a Household

Age dependency ratios are not only a measure of the dependents in a population, but are also a measure of the age structure of the population. Age dependency ratios are not only a measure of the dependants in a population but are also a measure of the age structure of the population. They relate to the number of individuals likely to be dependent on the support of others for their daily living – youths and the elderly – to the number of those individuals who are capable of providing such support. In all OECD countries higher life expectancy and lower fertility rates have led to a rise in the old-age dependency ratio and in a decline in the youth dependency ratio over time, although to very different levels and with various degrees of intensity in this decline (http:// www.oecd.org/dataoecd/4/24/38148786.pdf). Age dependency ratios indicate lower work participation and economic and social dependency on the young population, who are economically active. As the age of the persons who enter and move out of the workforce may vary, we have documented the non-working population: the students, retired from service, and the unemployed youth. This would reflect the number of persons in a household who are not productive because of various reasons: old age, students seeking education, unavailability of employment, and the like.

Basic Livelihood Needs

Domestic environment also contributes to meeting basic livelihood needs. Domestic environment includes housing, water supply, access to safe drinking water and use of fuel which indicates levels of poverty and development and is closely related to the health of women and children (Gopalan and Saksena, 1999).

Cooking and lighting – the two basic necessities – constitute 75 per cent of total energy consumed by rural population (Rajvanshi, 2005). The type of fuel used by a household speaks of the quality of life of the rural poor.

Women and girls are involved in two major household activities for survival: fetching water and gathering of fuel. The collection activity for both involves a lot of their energies and time in the task. They also suffer from various health problems arising from indoor air pollution. Most of their time is spent around the *chullah*. Emissions from traditional fuels like stoves for cooking contain very harmful pollutants. The use of fuel is closely associated with social, economic and health status of women and children (ibid., 1999).

It has been estimated that 80 per cent of all diseases in the world are associated with water usage. Women as primary users of water and sanitation facilities are the ones who are most affected by the lack or inadequacy of these facilities (ibid., 1999).

Household Assets

The data on income and consumption is not always reliable. Data on asset holding is an alternative to measure poverty or wealth (IFAD, 1992).

Bollom (1998) discusses consumer durables as an indicator measuring the level of wealth in a watershed area. He says, "The consumer durables are highly and easily visible and are difficult to conceal. Indirectly this indicator is also a measure of soil and water conservation". The consumer goods surveyed are selected with local culture and levels of wealth in mind.

Production and Income-Generating Activities

It is important to understand the household farm production and non-farm income-generating activities for their implications not only on poverty reduction and economic growth but also on the nutritional status of children.

Livelihood/Food Security

With watershed development there has been an increase in land-based livelihoods activities. It is important to understand whether the increased livelihoods have been able to support the family's food requirements. Migration, period of food shortage faced by the family and the loans undertaken give an idea of the alternative means sought by the families to supplement their livelihood as these have implications on the nutritional status of children.

The household schedule gathered household level information of children aged 0–5 years. The objective of this schedule was to gather information from the point of view of socio-economic status, status of drinking water in the families to which the children under study belonged.

The information was gathered on the following points:

(i) Background factors: Religion, caste, type of family.

(ii) Watershed development: Beneficiary/not, the kind of benefits the household has obtained from watershed development in the village.

(iii) Family profile: Total number of family members present and their background information, viz., sex and age composition, relationship with the head of the household, education, occupation, marital status.

(iv) Housing: type of house, cattle shed, location of cattle shed, presence of separate kitchen, its type and location.

(v) Type of fuel used.

(vi) Drinking water: Source, location, distance covered for fetching water, adoption of purification methods, water shortage period and duration.

(vii) Assets: Agriculture, kitchen, transportation, entertainment/educational.

(viii) Agriculture: Land holding, type, sources of irrigation, land under cultivation, crops taken, total surplus production, income, benefits obtained after watershed development, land use pattern: horticulture, floriculture productions, kitchen garden and the type of crops grown.

(ix) Livestock: Livestock holding, status of sale and purchase of animals in a year.

(x) Agriculture allied business: Dairy, poultry, sheep rearing, sale of animals and income from each, household consumption of milk, meat, eggs.

(xi) Distress migration, period, place of migration, reasons.

(xii) Food shortage: Period, reason.

(xiii) Loans taken in one year: Amount, duration, rate of interest, repayment.

Village Profile

Village settlements are of two types. Some are single settlement villages, others are nucleated. Smaller villages have a population of 200 persons or less, while the larger villages are inhabited by 5,000 people or more. The society is caste structured and the economy is mainly agricultural.

As already stated, the village profile schedule was developed to obtain information on the infrastructure available in the study villages which helped to understand the status of development therein. Data on the following parameters were collected:

(i) The number of hamlets into which the village is divided. The distance of each from the central village (*gaothan*).

(ii) The distance of the village from tehsil and district places. The connectivity of the village with the developed areas.

(iii) Availability and accessibility of the village with development services: education: schools and colleges, *anganwadi;* health care services: primary health centre and sub-centres, private clinics, number of visits paid by private doctors; traditional healing services: traditional birth attendants (TBA), trained and untrained, herbal specialists and bone setters, massagists, veterinary services; transport (public and private and their frequency) and communication services (post and telegraph, telephone, computer and internet); bank and cooperative societies, dairy cooperatives and private dairies.

(iv) Public distribution system (PDS): ration shops.

(v) Grain and seed banks.

(vi) Shops: medical, grocery and others.

(vii) Bakery, hotels.

If any of the above services are not available within the village, the nearest place where they are located and the distance, are taken into consideration.

Mother's Profile

The health of a child depends largely on the health of his/her mother. The health status of a woman since the early stages of her life until the stage of child bearing shapes the nutritional and health status of the child. The nutritional and health status of a woman since the early stages depends on her status and role in the family and the society. Sex ratio, age at marriage, age at first pregnancy, mobility, exposure, entrepreneurial activities and institutional membership are some of the areas that give an idea of the status of a woman. The present study generated a short profile of woman. The purpose was to have a quick understanding of the current status of woman in rural society.

Children's Schedule

The children's schedule focused on collecting data on practices related to child health influencing nutritional status. It contains questions regarding feeding practices, frequency of feeding, the food products given and changes in feeding practices.

Apart from the above discussed four main schedules, two more schedules were developed. These were anthropometry schedule and clinical examination schedule. While the schedule on anthropometry recorded different types of body measurements of children of age group 0–5 years under study, the schedule for clinical examination was developed to have a clinical assessment of child health pertaining to nutritional deficiencies and protein energy malnutrition (PEM) by a physician, and to provide medical advice and guidance to the community for improvements in the nutritional status of children. It was a small-scale intervention and study activity.

Training of Field Investigators

The training of the field investigators for data collection is an important element in the study, particularly more because of the involvement of village people in the data collection. A very focused training of the field investigators was carried out. Rather than focusing on the theoretical aspects of child nutrition, the training was geared towards how to elicit true information from the people in a gentle and modest way without being harsh and without causing harm to the informant.

It was a participatory training programme conducted for full four days. This consisted of an in-house training and practice session in the field for two days each. During the training, the students were imparted knowledge and skills in varied topics as follows:

(i) *Study topic:* Watershed development and child nutrition aspects.

(ii) *Rapport establishment:* Respect for the people, attitude of listening, learning and not of teaching, soft tone of conversation, being one with the people, building relationships with people, taking informed consent and its importance, dressing, grooming behaviour, participating in daily routine tasks of the people to help them complete the tasks, but non participation in aspects like local fights, household problems, engaging in talking to field village girls without reason.

(iii) Taking informed consent and maintaining confidentiality of data, the survey forms and the identity of people.

(iv) *Interview techniques:* Interviewing as a dialogue, recording of the data on survey forms. Avoiding of leading questions.

(v) *Anthropometry:* Five types of body measurements to be carried out in the study.

After the classroom training field practice was conducted in village Kalamkarwadi, a watershed village from Parner taluka, Ahmednagar. The field investigators were given interview schedules and assigned the task of data collection in the village. They were closely observed for the aspects they were trained. In the evenings meetings were organised with the field investigators to discuss their issues, practicality of interview schedules and to give feedback to them regarding their approach and work.

Village Meeting

An introductory meeting was organised in every study village before data collection. The purpose of conducting these meetings was to introduce the research work to be carried out in the villages and also the research team and request village level consent for the study.

The village meetings were planned well in advance. The senior staff of WOTR, Ahmednagar helped in coordinating with the village people to fix up the dates and timings for the meetings. In watershed villages, an official letter of request for the meeting was sent to the presidents of the community-based organisations like village watershed committee (VWC), the self-help group (SHG), the Samyukta Mahila Samiti (SMS) and other local leaders. Within the village, the meetings were communicated to all the people by using public address system such as that of temple loudspeakers and putting study information on the common notice boards of the villages at prominent places.

People generally preferred early morning meetings before they left to work in their farms. A group of villages, in close proximity were covered on the same day.

There have been different kinds of experiences in different villages. In some villages, both men and women participated in the meeting. In others, only women were present along with their children. In few villages, people did not gather for the meeting on the scheduled time. In such cases we went from house to house and requested them to gather for the meeting. In some cases the people did come. We however tried to reach out to people from the whole village and introduced ourselves and the purpose of study. Whatever the experience, efforts were taken to reach out to the grass roots. These meetings were attended by village men and women, the local leader, *sarpanch* and his associate, village level worker, the *gram sevak*, the gram panchayat members, *anganwadi* teacher, primary school teachers and members of village watershed committee and women's groups (SHG and SMS). In all the villages, people extended their wholehearted support for the study. They asked queries and gave suggestions.

During these meetings, a verbal consent was obtained from the villagers regarding conducting interviews, discussions, observations, photography, anthropometry camps for children and any other activity supporting the study. The accommodation arrangements for the study team were also discussed and the data collection was planned out.

The village meetings were conducted at common places in the village such as the *anganwadi* centre, village temple or the SHG office.

Data Collection

The data collection was carried out during December 2007 to July 2008 in the study villages.

In most of the villages the field investigators stayed in the village itself with a family. The accommodation and food arrangements were made by the people, the *anganwadi* workers, the *sarpanch*, small hotel owner's house and the health volunteers. They took appointments with the households, where data were to be collected. They could attend to early morning or late evening appointments. In few villages people were not able to arrange accommodation because of the agricultural workload. In such cases, the investigators stayed at places like hotel at taluka place and guest house of Sampada Trust at Ahmednagar and travelled to the villages for the purpose.

The survey work entails a lot of data management at different levels. At local level, to ease out management, each investigator was assigned a separate task such as maintaining daily attendance record, keeping track of the blank interview schedules, ensuring confidentiality of the filled up interview schedules, making local boarding and lodging arrangements, submitting the data to the office, etc. As part of the household data collection was related to women, the mothers of children, and as a section of the schedule had some sensitive questions, it was decided that one male and one female investigators should work in pairs.

In each village, one investigator was assigned the task of collecting data on village profile schedule in rotation. During the course of data collection, it was realised that the data collection for village profile needed investigators who are not having other tasks to perform. This was because information on village profile was to be gathered from several key informants such as the *anganwadi* teacher, SHG members, government officials such as *gramsevak, talathi* (official who maintains land records), etc. The availability of government officials was a constraint and several visits had to be made to cover their interviews. So, another two students of Rural Chroniclers course, one male and one female, were trained to work exclusively on the data collection of village profiles.

As stated earlier, the interviews began by taking an informed consent from the people. The informed consent form was read out to the informants and the study details were explained. For taking the anthropometric measurements of the children, the consent was taken from their mothers.

The forms were signed by the investigators as a confirmation of carrying out the consent process. The interviews were taken only where consent was obtained.

Monitoring of Data Collection

Monitoring of data collection is an important activity that helps to ensure and maintain the quality (reliability and accuracy) of the data. In the present study, monitoring of data collection was done in ten villages. More time was spent in the initial stages of data collection. The monitoring work was carried out by the researchers: the principal and the associates. The researchers followed two methods to monitor the data collection process. One was accompanying the investigator and observing the activities being conducted on the spot. The other was going back to the households in the absence of the field investigator and finding out from the member of the household about the activities conducted by him. These methods were followed only to ensure the quality of data. The following activities were conducted on the spot:

(i) Approaching the household
(ii) Initiating dialogue
(iii) Taking consent
(iv) Interviewing
(v) Anthropometry camp
(vi) Recording of the information on schedules
(vii) Concluding the interview

During the field monitoring, filled-in interview schedules were checked for data consistencies and to verify if all the data was correctly and properly filled as per the instructions given to the investigators.

In the late evenings, we conducted feedback training for the field investigators wherein each one shared their field experiences, problems, feedback of monitoring observations and discussions were done on varied issues to strengthen their capacities further in data collection. In the initial days of data collection the field investigators faced some difficulties. They were not well organised. Monitoring and feedback training helped them in improving their skills. A few observations are shared as follows:

It was observed that the field investigators had built up good rapport with the people. One experience we witnessed in this regard is that in one of the villages where the woman investigator stayed with the *anganwadi* worker's family, they treated her as their daughter. They presented her a *saree* and a blouse when she left to the next village, as is the tradition of sending off a daughter. The family also fell into tears when she was leaving.

During the dialogue, the field investigators tried to give some advice to the parents to take care of their children, and give substantial feeds. The issue was addressed in the feedback training. The role of the investigator as that of a learner in the study was emphasised.

In anthropometry camps, the steps while taking the measurements were observed and the discrepancies found were discussed with the investigators in the feedback training.

We also faced some difficulties in the first anthropometry camp organised in village Mhaswandi. As it was the first experience of having a camp for the field investigators as well as the researchers, we were not able to organise and manage it in a systematic manner. Therefore, the participation of children was poor. The investigators spent more time on filling information in the form for which children were not required. Even the smaller children cried while taking the measurements as they were scared by the instruments and the new faces of the investigators.

The papers were not kept ready to record the measurements of the child. The parents had to keep waiting for long. We realised that the investigators were well trained for taking measurements but were not able to manage a large number of children in the camp. In the feedback training we streamlined the process by allotting specific tasks in anthropometry camp to each investigator and ensured that the preparations were made in advance. Planning out the camp, taking measurements and recording the readings in the form with accuracy, managing the crying children and distribution of sweets at the end were the smaller tasks that were worked out and each investigator was assigned the specific task.

Based on these experiences, the investigators were then given feedback and instructed to make the required changes in organisation of the future camps. The camps in the subsequent villages were then conducted in a systematic manner.

Assessing Nutritional Status of Children

Indira Gandhi National Open University School of Continuing Education (IGNOU, year not known) has simplified and presented in detail the four major methods used to assess the nutritional status of individuals and population groups which include:

(i) Anthropometric measurement
(ii) Clinical method
(iii) Biochemical analysis
(iv) Diet survey

The present study uses anthropometric measurements to assess the nutritional status of children aged 0–5 years as an impact indicator of integrated watershed development and participatory growth monitoring activity.

Anthropometric Measurements

Although the study involves five anthropometric measurements for the analysis of the present work, the results of only two measurements have been discussed. Anthropometric indices are combination of measurements. They are essential for the interpretation of measurements. It is evident that a value for body weight alone has no meaning unless it is related to an individual's age or height. Thus, for example, measurements of weight and height may be combined to produce the body mass index.

Weight-for-Age

Weight-for-age reflects body mass relative to chronological age. Low weight-for-age reflects stunting or wasting or both, i.e., long-term and short-term effects of undernourishment. However, it cannot differentiate between present (acute) and past (chronic) malnutrition. 'Lightness' has been proposed as the descriptive term for low weight-for-age, while 'underweight' has been used to refer to the underlying pathological process (WHO Expert Committee Report, 1995).

Weight is the most common and simplest anthropometric measurement used for the assessment of the nutritional status of children. It is an indicator of the body mass and is a composite of all body constituents like water, minerals, fat, protein, bone, etc. It is sensitive to even small changes in the nutritional status due to childhood morbidities like diarrhoea, etc. (Rao Hanumantha and Vijayaraghavan, 1996). The increase in weight indicates gain in body tissue and is a feature of growth. Loss in body weight means wasting or loss of body tissue (mostly fat or muscle) (Bhattacharyya, 2000).

Height-for-Age

"Height is one of the important anthropometric measurements. It is a linear measurement and sum of four components, that is, legs, pelvis, spinal cord and skull. Height of a person is determined by genetic (hereditary) and environment factors. In the environmental factors the most important are nutrition and morbidity. During periods of severe deprivation, linear growth rate slows down and leads to short stature in an individual" (Rao Hanumantha and Vijayaraghavan, 1996).

"The increase in height takes more time to be obvious than gain in weight and height may remain stationary when weight is either stationary or being lost" (Bhattacharyya, 2000).

Height-for-age reflects achieved linear growth and its deficits indicate long-term, cumulative inadequacies of health or nutrition.

Low height-for-age is also called as shortness or stunting. Shortness is the descriptive definition of low height-for-age. A given deficit in height may represent a short period of growth failure at an early age or a longer period of growth failure at a later age (Gopaldas and Seshadri, 1987).

For the present study, this index will be used primarily along with other indicators as it best measures the long-term effects of undernutrition and is one of the indicators highly recommended by Bollom (1998). He suggests that height-for-age is an indicator of wealth for very poor. It is indirectly a measure of soil and water conservation in a rural community. Improved health is often linked to the raw materials of farming.

Anthropometric measurements are in themselves of no use unless they are used in relation with other variables for proper interpretation. These indices are used as the main criteria for assessing the adequacy of diet and growth in infancy (WHO Expert Committee Report, 1995).

Measurement Techniques and Steps

Weight Measurement

Instrument

Baby weighing scale was used for infants who were unable to sit on their own. The scale was 'Detecto Baby scale', manufactured by De National Scale Company, Shahpur, India. It had a tray where the infant could be lied down for taking the measurement and the maximum weight that could be measured by it was 10 kg.

Weight of children who could stand on their own and whose weight was more than 10 kg were weighed using platform weighing scale called as 'Fitnesse, Libra Personal Scale' manufactured by Edryl India.

Preparation

Before weighing, the investigators adjusted the scale to zero reading and placed it on a plain level surface. It was ensured that the children had minimum clothing and were without any foot wears during weighing. Parents were of great help in this regard. Earlior, it was decided to weigh children in inner clothing with the help of parents. But, they did not like it as they were worried and felt unsafe taking out their outer clothing. Therefore, we took the measurement in clothes.

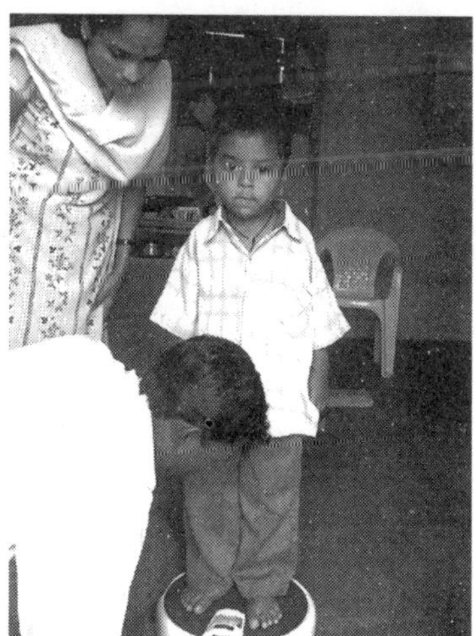

Plate 6: Investigators measuring weight of a child using bathroom scale

As a general precaution, the weighing scales were always kept inside cover when not in use.

Technique

The child was asked to stand erect on the weighing scale without taking the support of anything and looking straight ahead. The measurer took the reading while the recorder noted down the measurement on the anthropometric form. For weighing infants, they were made to lie down on the tray of baby scale and then the reading was taken. However, at times this was difficult as the baby would cry and make constant body movements resulting in fluctuations in the readings. Hence, it was ensured that the pointer of the scale was steady and only then the reading was taken.

In some cases, children refused to stand on the weighing scale, as they felt scared. In such conditions, the accompanying parent was first weighed alone and the reading noted down. Then the same person was weighed with the child and the reading was taken. The weight of the child was then calculated by subtracting the first reading from the second. Plate 3 depicts the weight of a child being taken by investigators using platform-weighing scale (bathroom scale).

Height Measurement

Two related terms – length and stature – are also used for height. Length represents the measurement taken in a recumbent position and is often used for children 2–3 years of age who cannot stand well. Standing height measurement is referred to as stature (WHO Expert Committee Report, 1995).

Instrument

For measurement of standing height, height stands manufactured by Hardic Meditech, Delhi, India were used. The stand consisted of three detachable parts. This made it easy to carry in the field. There was a platform for standing and two rods, one of which had a movable arm for taking measurement.

In case of children who could not stand and measured less than 90 cm, recumbent length was measured using infantometer. It consisted of a fixed head board and a movable leg board.

Preparation

Before taking measurement by the height stand, all the three parts of stand were fixed together. As far as possible, it was placed on a level ground and against a solid and flat vertical surface. In case of infantometer it was kept on a plain surface. It was ensured that the child did not wear footwears and head covers.

Technique for Measuring Recumbent Length

The technique requires two investigators to take the measurement. The baby was laid down on the infantometer gently with the head towards the fixed headboard. The body was straight in line with the board, with the head, back and buttocks touching it. One of the investigators held the head of the baby against the headboard with the baby looking straight up. The knees of the baby were pressed down gently so as to extend the legs as fully as possible. The sliding legboard was then moved to touch the feet of the baby with his toe pointing upwards. At this point of time the reading was taken and recorded by another investigator.

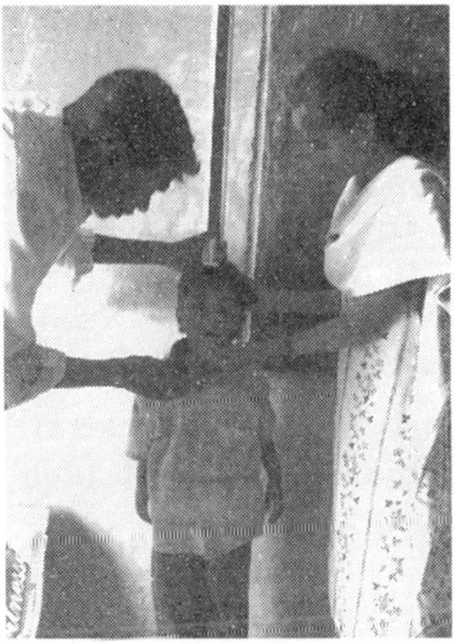

Plate 7: Investigators measuring height of a child using height stand

However, taking this measurement was a bit difficult in view of the age of children. However, with the help of the parent or the care giver who accompanied the child the measurement was taken.

Technique for Measuring Standing Height

The child was made to stand erect on the standing platform with feet running parallel to each other and looked straight ahead. The heels touched each other with the toes apart, arms at side and the palms touching the thighs. The heels, back, neck and the head touch the stand. The moving arm of the board was then slightly moved down to touch the child's vertex. A very slight pressure was applied to reduce the thickness of the hairs. At this point of time, the reading was taken by the investigator and recorded by another investigator on the form.

Reference Standards in Anthropometry

For evaluating and interpretation of anthropometric measurements of an individual, it has to be compared with a set of values. These values are called as 'growth standards' or references. The anthropometric status of an individual measurements obtained on well-to-do children of elite parents are usually treated as 'reference standards'. The present study makes use of reference standard of National Institute of Nutrition, Hyderabad, complied by Ghosh and Shanti in 1986.

Classification System for Anthropometric Indices

There are different types of classifications available to assess the nutritional status of children based on weight-for-age and height-for-age. In the present study, for assessment based on weight-for-age, Gomez classification has been used whereas for assessment based on height-for-age, Watorlow's classification has been used. These classifications are widely used and preferred.

Anthropometry Camps

In every village, anthropometry camps were organised after the completion of the household-level interviews. The preparations for the camps were made well in advance. The camps were held in places like *anganwadi* centre in most of the cases, office of women's committee (SMS) and village temple. The dates and timings of the camps were decided in consultation with the people and the *anganwadi* workers. The parents of the children were informed about the time schedule about one or two days earlier. The timings of the camps were fixed in such a manner that it would be convenient to the people to bring their children and that it would not interfere with the routine work of the *anganwadi* centres. As people generally preferred morning timings before going to the farms, the camps were generally held during 8 to 10 a.m.

Home visits were made to take the measurements of children who were not able to participate in the camps. In large villages having more than one *anganwadi* centre, the camp was organised at different places in the village convenient for the children to participate.

As the study involved taking body measurements and assessing nutritional status of children, initially the *anganwadi* workers feared and misunderstood the activity as an evaluation of their work. It was explained to them how it was a part of a research study and that the purpose of the study was not to evaluate *anganwadi* work but was actually a part of an evaluation of watershed development programmes undertaken by an NGO. The *anganwadi* workers extended all support and cooperation required in the camps.

Since the measurements pertained to children of 0–5 years of age, managing and taking proper readings was a challenging task. In all the cases, two investigators took the readings and another investigator recorded them on the forms. The remaining investigators helped in managing the crowd and maintaining silence.

Diet Survey Using 'Twenty-Four Hours Recall' Method

Information relating to the food consumed by the children one day before the survey was collected using 'twenty-four hour recall' method. The idea of using this method in the present study was to understand the food intake of children across the comparative groups rather than to assess the nutritional status. Diet is a vital determinant of health and nutritional status of people.

In this method, exact amount of food consumed in terms of grams and litres was assessed and their protein and energy content was estimated. A comparison of food intake was carried out with the Recommended Dietary Allowances (RDA) for analysis as developed by ICMR-NIN standard. RDA provides a measure of adequacy or inadequacy of food/nutrient intake.

In this diet survey method the parent/care giver was asked about the quantity of food consumed by the children on the previous day of the survey. The consumption was recorded either in form of raw ingredients that were used for cooking or in form of cooked food consumed. We could not stick to either of the consumption in all households because of practical difficulties. Most informants were able to tell the quantity of food consumed by the child in terms of cooked food. Based on the food consumed by the child during the whole day, the energy and protein consumption was then calculated.

Standard measuring cups and jars were used to estimate the amount of food consumed by the children. The cups were shown to the mother/care giver of the children and they would then tell the quantity of consumption of the child. Consumption of major food ingredients of any particular food item was recorded. Minor food ingredients such as salt, spices were not considered for analysis.

In general, the method got executed well in the field but there were some practical difficulties that we would like to share and are given below:

(a) In some cases, the mother could not recall easily the consumption of the child. In such cases, she was given more time to recall properly.

(b) At times, mother of the child was not aware of food consumption at a particular time as she was out for some work or at the fields. In such cases, information was also sought from the care givers who were usually the elderly persons such as grandmother.

(c) At times, the mother/care giver did not pay attention while the child ate and hence did not know the exact amount consumed by the child, out of the total quantity that was given to him for eating.

(d) The quantity of *anganwadi* food consumed by the child was based on the best assumption that the mother/care giver could give as the anganwadi food is generally consumed by the children in the anganwadi itself. So, the mother told the quantity of the food that the child generally consumes.

(e) Reporting of the liquid food items was relatively simple for the informants.

Calculation of Nutrient Contents of Food Items Consumed by the Children

For calculating the nutrient contents of various food items consumed by a particular child, first data on actual consumption of food by the child on the previous day was obtained using 24–hour diet recall method. As stated earlier, while recording the consumption data, it was recorded that of the cooked food for most of the items as reported by the mother/care taker.

For calculating the energy and protein contents of raw food ingredients, values provided by National Institute of Nutrition (NIN) have been used (reference book: *Nutritive Values of Indian Foods,* NIN, Hyderabad).

For calculating the energy and protein contents of cooked food ingredients, values provided by Agharkar Research Institute, Pune, were used (reference book: *Nutritive Values of Some Indian Food Preparations,* Agharkar Research Institute, Pune).

Immunisation of Children

The United Nations' Millennium Development Goals aim at reducing the under-five child mortality rate by two thirds by the year 2015. Vaccination of children for the life-threatening diseases is one of the essential ways towards reaching this goal. Immunisation protects children from six vaccine-preventable diseases – poliomyelitis, diphtheria, pertussis, tetanus, tuberculosis and measles. These are major preventable causes of child mortality, disability, morbidity and related malnutrition (www.wcd.nic.in). The Ministry of Health and Family Welfare provides the immunisation services through its Reproductive Child Health (RCH) programme. In addition to the above vaccines, Vitamin 'A' and iron supplementation of children is also done under this programme.

According to the guidelines developed by World Health Organisation (WHO), a child is considered fully vaccinated when he/she has received three doses each of polio and DPT and one dose each of tuberculosis (BCG) and measles. According to the immunisation schedule outlined by Government of India and the WHO, all primary vaccines should be administered to the child by the age of 12 months (IIPS, 2006).

In the present study, all children who received three doses each of polio and DPT and one dose each of tuberculosis (BCG) and measles are considered fully vaccinated.

Data Entry and Quality Assurance

The task of data entry of the study was outsourced. It was carried out using the SPSS software (version 14). The questions in the interview schedules were first coded and then the data entry was carried out. For ensuring quality in the data entry process, 10 per cent of the total data entry was verified by the data entry personnels and also the researchers. Some of the schedules were rechecked for verifications and some of the sections had consistent errors. The complete data entry of these sections was carried out again so as to ensure least possible errors in the data entry. This re-entered data was again verified by the researchers and only when it was found to be error-free, the further data analysis was carried out by the statisticians.

Data Analysis

Analytical frameworks were developed to help the statisticians to conduct analysis in line with the objectives of the study. Regular meetings with the statisticians were conducted to acquaint them with the various analytical frameworks.

Data Interpretation

The results obtained from the statisticians were in the form of huge tables. Management of the tables itself was a great and

Figure 3.6

Research steps

Preparation of concept note
↓
Conducting the review meetings
↓
Development of research proposal
↓
Discussion and refinement of research proposal
↓
Selection of study sites
↓
Sample selection
↓
Development of tools of data collection
↓
Review meetings and discussion regarding tools
↓
Refinement of tools of data collection
↓
Selection of field investigators
↓
Training of researchers
↓
Training of field investigators
↓
Conducting village meetings
↓
Taking informed consent
↓
Data collection, data monitoring and feedback trainings (simultaneous activities)
↓
Data coding
↓
Data entry
↓
Data analysis
↓
Interpretation of data
↓
Report writing

challenging task. Each table was broken down and simplified by the researchers. This was a very tedious process or in simple words, it was a kind of surgery that was carried out with the tables. The tables were simplified to make it readable and meaningful. The whole process was carried out in consultation with the statisticians. Statistical analysis becomes more meaningful if the results are interpreted. The interpretation of the results has been discussed out and is presented in Chapter 8 of the book.

The research process that was carried out in the study is briefed out in the form of flow chart as shown in Figure 3.6. It explains the various steps that were carried out to develop the study. It may be pertinent to mention that the review of literature was an ongoing activity, starting from concept note preparation to report writing.

The results of the study have been presented in chapters 4 to 7. The numerical data is presented in tabular form across the comparative study groups like the villages with watershed development and child growth monitoring inputs, villages with watershed development and no child growth monitoring and villages with no such inputs.

Chapter 8 discusses the analysis and interpretation of the data presented in the above mentioned chapters.

Chapter 9 discusses how watershed development can give direction for improving the health status of the community. As this study is based on a hypothesis, the validity of the hypothesis has also been discussed.

4

Anthropometry, Child Health, Feeding and Dietary Practices

In this chapter, data on nutritional status of children aged 0–5 years have been presented. The nutritional status is based on anthropometric indices, immunisation status and feeding and dietary practices.

Anthropometry

Height-for-Age: The Impact Indicator

Status of Total Population

The results indicate that the overall nutritional status of children in the region is fine, that is, more than 60 per cent of the children are normal for their height-for-age.

The percentage of children, irrespective of caste/tribe having normal nutritional status, are slightly more in No Input group in terms of height-for-age, followed by WSD+GM and WSD groups. There is not much difference between No Input and WSD+GM groups (0.9%). When the watershed villages are compared, the percentage of children having normal nutritional status is more in WSD+GM group than in WSD group and the difference is of 1.8 per cent.

Table 4.1

Percentages of normal children across comparative study groups

Comparative group	*Community**	*Sample size*	*Height-for-age*		*Weight-for-age*	
			Number	*Percentage*	*Number*	*Percentage*
WSD+GM	Tribe	16	9	56.3	7	43.8
	Upper	74	50	67.6	33	44.6
	Lower	12	5	41.7	4	33.3
	Other	4	2	50.0	0	0.0
	Overall	106	66	62.3	44	41.5
WSD	Tribe	31	13	41.9	10	32.3
	Upper	70	44	62.9	40	57.1
	Lower	18	13	72.2	12	66.7
	Other	5	5	100	5	100
	Overall	124	75	60.5	67	54.0
No Input	Tribe	17	8	47.1	7	41.2
	Upper	83	57	68.7	33	39.8
	Lower	10	5	50.0	3	30.0
	Other	4	2	50.0	2	50.0
	Overall	114	72	63.2	45	39.5
Total		344	213	61.9	156	45.3

* Please refer to the Appendices to understand the communities included in different categories.

Status Across Communities

Tribal Community

The percentage of children having normal nutritional status in tribal community is more in WSD+GM group, followed by No Input and WSD groups in terms of height-for-age.

There is a vast difference between WSD+GM and the other two groups. The WSD+GM group is much ahead of No Input group (the difference being 9.2%) whereas the No Input group is ahead of WSD group (the difference being 5.2%). The difference between WSD+GM group and WSD group is of 14.4 per cent. This indicates that the inputs of growth monitoring play an important role in the nutritional status of children belonging to the tribal community.

Upper/Dominant Caste Community

The percentage of children having normal nutritional status in upper caste community are more in No Input group, followed by WSD+GM and WSD groups, in terms of height-for-age. The difference between No Input and WSD+GM group is small (1.1%), whereas the difference between No Input and WSD group is of 5.8 per cent and WSD+GM group is ahead of WSD group by 4.7 per cent.

Lower Caste Community

The percentage of children having normal nutritional status in lower caste community is more in WSD group, followed by No Input and WSD+GM groups, in terms of height-for-age. The difference between WSD and other groups is vast in this regard. The WSD group is ahead of No Input group by 22.2 per cent and WSD+GM group by 30.5 per cent.

Other Communities

The percentage of children having normal nutritional status in other communities is 50 more in WSD group than other groups.

Weight-for-Age

Status of Total Population

The results indicate that the overall nutritional status of children in the region is poor as about 55 per cent of the children are not normal as per weight-for-age index.

The percentage of children, irrespective of caste/tribe and having normal nutritional status, is more in WSD group in terms of weight-for-age, followed by WSD+GM and No Input groups. There is a large difference between WSD and other groups. WSD group is ahead of WSD+GM group by 12.5 per cent and No Input group by 14.5 per cent. The difference between WSD+GM and No Input group is less (i.e., 2%).

Status Across Communities

Tribal Community

The percentage of children having normal nutritional status from tribal community is more in WSD+GM group, followed by No Input and WSD groups in terms of weight-for-age. The difference between WSD+GM and No Input group is not much (2.6%) whereas between WSD+GM and WSD group it is vast (11.5%). This indicates that the inputs of growth monitoring play an important role in the nutritional status of children belonging to the tribal community.

Upper/Dominant Caste Community

The percentage of children having normal nutritional status in upper caste community is more in WSD group, followed by WSD+GM and No Input groups, in terms of weight-for-age. The difference between WSD and other groups is vast. While WSD group is ahead of WSD+GM group by 12.5 per cent and No Input group by 17.3 per cent among the children belonging to upper caste, the WSD+GM group is ahead of No Input group by 4.8 per cent.

Lower Caste Community

The percentage of children having normal nutritional status in lower caste community is more in WSD group, followed by WSD+GM and No Input groups, in terms of weight-for-age. The difference between WSD and other groups is vast in this regard.

The WSD group is ahead of WSD+GM group by 33.4 per cent and No Input group by 36.7 per cent whereas the WSD+GM group is ahead of No Input group by 3.3 per cent.

Other Communities

The percentage of children having normal nutritional status in other communities is 50 per cent more in WSD group than in No Input group and 100 per cent more than WSD+GM group groups based on weight-for-age criteria.

Ranking Nutritional Status of Children in Three Groups

For a simple and easier understanding, the results are given in ranks of 1, 2 and 3. The comparative group where most of the children are having normal nutritional status according to an anthropometric index are assigned rank 1 and a group with the least number of children having normal nutritional status is assigned rank 3. Children in between are ranked 2. Based on these ranks, the following table has been derived:

Table 4.2

Ranking of the comparative groups based on the nutritional status of children

Comparative group	*Height-for-age*	*Weight-for-age*
WSD+GM	2	2
WSD	3	1
No Input	1	3

Child Health, Feeding and Dietary Practices

In total, 354 children of age 0–5 years have been covered in the survey that collected data on children's schedule, out of which 193 are boys and 161 are girls. The distribution of the sample size across comparative groups is 111 children from WSD+GM group, 127 from WSD group and 116 from No Input group.

Average Birth Weight

The average birth weight of children in WSD+GM, WSD and No Input groups is 2.59 kg 2.68 kg. and 2.74 kg respectively.

Feeding Practices

Breast-feeding

The initiation of breast-feeding has taken place three days after delivery in 67 per cent of the children and the colostrum is discarded. There is not much difference across comparative groups in this practice as all follow the same traditional practice. It is believed that the breast milk is not initiated until three days after delivery and colostrum is believed to be hard to digest for the new-born baby. Illness of mother was also highlighted as one of the reasons for not initiating breast-feed. Instead, external feeds like sugar or jaggery water, glucose water, honey, cow/goat milk are given to the baby during the first three days after birth. The reasons cited for giving external feeds to their babies were that they felt hungry, they cried and that if external feeds are not given they would become weak.

In the remaining cases, wherein breast-feeding of babies was started by mothers immediately after delivery, it was done so on medical advice, followed by media influence and mother's self-awareness about the importance of colostrum feeding for enhancing child immunity and health.

Weaning

The average age of starting supplementary food in WSD+GM, WSD and No Input groups is 10.8, 10.5 and 12.2 months, respectively.

The average age of stopping breast-feeding in WSD+GM, WSD and No Input groups is 19.4, 15.7 and 19.4 months, respectively.

The introductory supplementary food items include *dal* and rice, biscuits, liquids such as *dal* water, fruit/fruit juice, *chapatti*, *bhakar*, rice *pej* (water of boiled rice), milk, rice and milk, and farex/cerelac. The most common food item is *dal* and rice. In 27 (25.5%), 49 (40.5%) and 34 (31.2%) of the children in WSD+GM, WSD and No Input groups respectively, the introductory supplementary food item is *dal* and rice. After dal and rice, the most common item is rice *pej* in all the three groups.

The most common reason for starting with the specific food item in all the three groups is its ability to digest easily. The second most common response is that the food item is nutritious.

The mothers who have not started giving supplementary food to their children reported that they would start it at an average age of 9.8, 9.8 and 7.5 months in WSD+GM, WSD and No Input groups, respectively.

Complementary Feeding

The number of children consuming milk other than breast milk aged six month onwards are 78 (70.3%), 79 (62.2%) and 78 (67.2%) in WSD+GM, WSD and No Input groups, respectively.

The first preference is cow/goat milk followed by buffalo milk. There is one case in village Mhaswandi (WSD+GM group village) where the mother reported giving milk product like milk powder or infant formula to tho child.

About 90 per cent of the children in all the three groups are fed external milk with the help of a container and spoon. The remaining feed with the help of bottle.

In more than 95 per cent of the cases, the milk is diluted with water before feeding it to the children. In case of cow's, goat's and buffalo's milk, the dilution ratio is 1:0.25.

Food Intake

Protein Energy Consumption

This section presents the average energy and protein consumption of the children for one day (in Kcal) and its comparison with the Recommended Dietary Allowance (RDA).

The children of age 0–1 year have been ignored for the above calculations. This is because they are mostly on breast-feed until at least six months of age and derive their energy and protein requirement from breast milk only. Even in children aged 7–12 months, breast milk constitutes major part of the day's diet as weaning takes place much later.

Table 4.3

Calculation of average energy consumption for one day (in Kcal) and its comparison with the Recommended Dietary Allowance (RDA)

(N: WSD+GM = 105, WSD = 123, No Input = 110)

Age group	*Category*	*WSD+GM*	*WSD*	*No Input*
1–3 years	Recommended energy consumption: 1240 Kcal			
	Actual energy consumption	572	484	698
	Percentage	46.1	39.0	56.3
4–5 years	Recommended energy consumption: 1690 Kcal			
	Actual energy consumption	794	679	1054
	Percentage	47.0	40.2	62.4

The above table shows that the actual energy consumption is low in the age group 1–3 years in all the three groups under study. It also indicates that the actual energy consumption is better in No Input group and of lowest value in WSD group in both the age groups, that is, 1–3 and 4–5 years.

Table 4.4

Calculation of average protein consumption for one day (in gm) and its comparison with the Recommended Dietary Allowance (RDA)

(N: WSD+GM = 105, WSD = 123, No Input = 110)

Age group	*Category*	*WSD+GM*	*WSD*	*No Input*
1–3 years	Recommended protein consumption: 22 gm			
	Actual protein consumption	18.8	17.2	23.3
	Percentage	85.5	78.2	105.9
4–5 years	Recommended protein consumption: 30 gm			
	Actual protein consumption	26.1	21.8	34.0
	Percentage	87.0	72.7	113.3

The table further reveals that the actual protein consumption is better than actual energy consumption in all three groups in both the age groups of children, that is, 1–3 and 4–5 years.

Besides, the actual protein consumption is much better in No Input group and lowest in WSD group.

It may be pertinent to mention that for Tables 4.3 and 4.4, the recommended consumption values are as recommended by Indian Council of Medical Research (ICMR). The percentages are calculated based on these recommended values.

Food Frequency

Food frequency was one of the tools developed to understand the consumption of specific food items. The results are presented below:

Fruits

(N: WSD+GM = 79, WSD = 91, No Input = 72)

The children from all the three studied groups consume fruits every week. Consumption of fruits at least twice a week is common. The numbers and percentages of such children are 28 (35.4%), 37 (40.7%) and 18 (25.0%) in WSD+GM, WSD and No Input groups, respectively. Thus, WSD group is much ahead in this regard, followed by WSD+GM and No Input groups.

Green Leafy Vegetables (GLV)

(N: WSD+GM = 70, WSD = 95, No Input = 78)

The children from all the three groups consume GLV every week. Consumption of GLV at least twice a week is a common practice. Their numbers and percentages are 35 (50.0%), 45 (47.4%) and 27 (34.6%) in WSD+GM, WSD and No Input groups, respectively.

The data thus indicate that more number of children from WSD+GM group consume GLV twice weekly followed by WSD and No Input group respectively. While there is not much difference between WSD+GM and WSD groups, the consumption of GLV is comparatively lower in No Input group.

Other Vegetables

(N: WSD+GM = 32, WSD = 59, No Input = 34)

The children also consume other vegetables at least once in a day in all the three groups under study. The numbers and percentages of such children are 32 (100.0%), 54 (91.5%) and 33 (97.1%) in WSD+GM, WSD and No Input groups, respectively. Thus, there are a fair number of children consuming other vegetables once in a day in all the three groups (that is, above 90%). However, they are more in WSD+GM group.

Legumes

(N: WSD+GM = 43, WSD = 37, No Input = 21)

The children from all the three groups consume legumes every week. Their numbers and percentages are 19 (44.2%), 26 (70.3%) and 9 (42.9%) in WSD+GM, WSD and No Input groups, respectively. Thus, there are more number of such children in WSD group, followed by WSD+GM and No Input groups. There is not much difference in the consumption of legumes in children between WSD+GM and No Input group but the frequency is found to be higher in WSD group setting it widely apart.

Pulses

(N: WSD+GM = 75, WSD = 88, No Input = 80)

The children from all the three groups consume pulses at least twice a week. Their numbers and percentages are 37 (49.3%), 31 (35.2%) and 32 (40.0%) in WSD+GM, WSD and No Input groups, respectively. Thus, there are more number of such children in WSD+GM group followed by No Input and WSD groups. The number of children is higher in WSD+GM group and lower in WSD group.

Roots and Tubers

(N: WSD+GM = 35, WSD = 22, No Input = 10)

The children from all the three studied groups consume roots and tubers at least twice a week. The numbers and percentages of such children are 19 (54.3%), 3 (13.6%) and 5 (50.0%) in WSD+GM, WSD and No Input groups respectively. Thus, there are more children consuming roots and tubers in WSD+GM group followed by No Input and WSD groups. There is not much difference between WSD+GM and No Input group in this respect but the frequency of consumption of roots and tubers is much lower in WSD group.

Porridge: Lapsi/Bharda

(N: WSD+GM = 53, WSD = 55, No Input = 63)

The porridge is made up of wheat and jaggery fried in *ghee.*

The children from all the three groups consume porridge at least twice a week. Such children constitute 32 (60.4%), 30 (54.5%) and 7 (11.1%) in WSD+GM, WSD and No Input groups, respectively.

Thus, there are more number of such children from WSD+GM group followed by WSD and No Input groups. The consumption of porridge is higher in WSD+GM group and lower in No Input group.

Yogurt

(N: WSD+GM = 23, WSD = 10, No Input = 6)

Consumption of yogurt twice a week is also common among children. The numbers are 11 (47.8%), 5 (50.0%) and 1 (16.7%) in WSD+GM, WSD and No Input groups respectively.

Baby Food (Farex, Cerelac)

The consumption of baby food such as Farex/Cerelac is also common among children in all the three groups. It is generally given twice a week.

Eggs

(N: WSD+GM = 39, WSD = 41, No Input = 36)

The children consume eggs at least twice every week. The numbers and percentages of such children are 13 (33.3%), 21 (51.2%) and 12 (33.3%) in WSD+GM, WSD and No Input groups, respectively. Thus, there is no difference in WSD+GM and No Input groups and the consumption is lower in both while the consumption of the eggs is higher in WSD group.

Fish

(N: WSD+GM = 9, WSD = 7, No Input = 3)

In all the three groups the children consume fish at least twice every month. The numbers and percentages of such children are 6 (66.7%), 3 (42.9%) and 1 (33.3%) in WSD+GM, WSD and No Input groups, respectively. Thus, more number of children from WSD+GM group consume fish followed by WSD and No Input groups. The consumption of fish is higher in WSD+GM group and lower in No Input group.

Chicken

(N: WSD+GM = 22, WSD = 35, No Input = 20)

Monthly consumption of chicken is more common in all the three groups as compared to daily or weekly consumption. The numbers and percentages of children consuming chicken twice a month are 6 (27.3%), 16 (45.7%) and 14 (70.0%) in WSD+GM, WSD and No Input groups, respectively.

Meat

(N: WSD+GM = 18, WSD = 27, No Input = 19)

The children from all the three studied groups consume meat at least twice a month. Their numbers and percentages are 8 (44.4%), 10 (37.0%) and 10 (52.6%) in WSD+GM, WSD and No Input groups, respectively. Thus, there are more children consuming meat in No Input group followed by WSD+GM and WSD groups. The consumption is comparatively lower in WSD group.

Food Insufficiency

The data on food insufficiency was collected by asking a question to the parent if the child ate food less than usual in terms of frequency of meals in one year preceding the survey, i.e., December 2006–November 2007. The usual pattern is that the children ages one year onward have food three to four times a day.

The total number of children eating less than usual in one year preceding the survey is 3 (2.7%), 0 (0.0%) and 2 (1.7%) in WSD+GM, WSD and No Input groups, respectively.

In WSD+GM group, the three children faced food insufficiency for a total of 103 days in one year during the months of March, April, May and October. The reason for eating less than usual is lack of agricultural work due to less rainfall.

In No Input group, the two children faced food insufficiency for a total of 75 days in one year during the months of May, June, July, August and October. The reasons were inability of the head of the household to work duo to illness and lack of agricultural work.

The figures indicate that eating less than usual is comparatively more in WSD+GM group, followed by No Input group and eating less than usual is not reported in WSD group.

Immunisation Status of Children

The study also looked into the coverage of basic/primary vaccination of children as per their age. The schedule of immunisation as per age developed by the ICDS and presented below (Table 4.5) is used as the reference standard to define the immunisation status of the children.

Table 4.5

Immunisation schedule for children (based on ICDS)

Vaccine	*Age*				
	Birth	*6 weeks (1.2 months)*	*10 weeks (2.2 months)*	*14 weeks (3.2 months)*	*9–12 months*
Primary Vaccination					
BCG	√				
Oral Polio	√	√	√	√	
DPT		√	√	√	
Measles					√

A child is considered as completely immunised if he/she has received three doses each of DPT and oral polio vaccines and one dose each of tuberculosis (BCG) and measles vaccines. The primary vaccines as mentioned in the table above are to be administered to the child by the age of 12 months. It is simpler to look into the status of complete immunisation for children who have completed and are older than a year. Because, by one year, the primary vaccination gets completed. For children who are of age up to 11 months, the status of complete immunisation would be defined on the basis of primary vaccines received as per age.

Table 4.6

Immunisation status of children

Immunisation status	*Category*	*Completely immunized*	*Partially immunised*	*Not received single vaccine*	*Missing*	*Total*
WSD+GM	Number	74	36	1	0	111
	Percentage	66.7	32.4	0.9	0	100
WSD	Number	104	22	0	1	127
	Percentage	81.9	17.3	0	0.	100
No Input	Number	89	26	1	0	116
	Percentage	76.7	22.4	0.9	0	100

The above table indicates that more number of children from WSD group are completely immunised followed by the No

Input and WSD+GM groups. The number of children who are partially immunised are more in WSD+GM group, followed by the No Input and the WSD groups. The percentage of children who have not received a single vaccine is lesser but when compared among the groups, the WSD group has no such child.

Consumption of Tonics and Gripe Water

Tonics

The number of children who were given tonics at any point of time since birth is 18 (16.2%), 6 (4.7%) and 12 (10.3%) in WSD+GM, WSD and No Input groups, respectively.

The figures indicate that more number of children from WSD+GM group have taken tonics since birth followed by No Input and WSD groups. Thus, the number of such children is comparatively lower in WSD group.

The tonics were given to the child on the advice of doctor, relative/friend and own knowledge. However, more number of children were given tonics on doctor's advice. The number of children given tonic on doctor's advice are 10 (55.6%), 4 (66.7%) and 4 (33.3%) in WSD+GM, WSD and No Input groups, respectively.

However, the following information presents a difference in the list of tonics given to children across comparative groups:

WSD+GM group: Corinite syrup, ferratola-12, ferrofade, himalt-x, lupactix, pro vitamin, probe granules, nutritional supplement, procal, rudimin, sofzid and vitamin.

WSD group: Bevon drops and FDICOL.

No Input group: AVIT, BITCO, Errotola-12, Febrex Plus, Wikozyl-Ozal, Ferrofola-12, Fersoft, Airstozyme and Pro vitamin.

The above information indicates that the number of tonics given is more in WSD+GM group followed by No Input group. It is lower in WSD group.

Gripe Water

The number of children who are given gripe water across the groups are 15 (13.5%), 5 (3.9%) and 25 (21.6%) in WSD+GM, WSD and No Input groups, respectively. The figures indicate that more number of children who are given gripe water are from No Input group followed by WSD+GM and WSD groups. Most of the children are given gripe water purchased from the market. None of the children are found to be given home-made gripe water in all the three groups. The gripe water is usually given twice a day.

The following reasons were given by the parent/care giver of the children for giving gripe water:

(i) To make baby sleep.
(ii) It acts as a medicine for gastric trouble.
(iii) It is good for baby's health.
(iv) It quietens the crying child.

In WSD and No Input groups more number of children are given gripe water because it is considered good for baby's health. The number of such children are 2 (40.0%) and 11 (44.0%) in WSD and No Input groups respectively. In WSD+GM group more number of children are given gripe water to make the baby quiet. The number of such children are 4 (26.7%).

Morbidity History

An attempt has also been made to take the morbidity history of the child since birth. It was found that people could tell the morbidity history of their child for five years. The maximum number of illness episodes reported is four. Children who fell ill four times since their birth are 0 (0.0%), 0 (0.0%) and 1 (0.9%) in WSD+GM, WSD and No Input groups, respectively. Children who fell ill at least once since their birth in WSD+GM, WSD and No Input groups are 78 (70.3%), 64 (50.4%) and 92 (79.3%), respectively.

The children from all the three groups have mostly suffered from common health problems like cough, cold and fever. The number of children suffered from water-borne diseases like gastro, diarrhoea and jaundice were not much.

Duration of Illness

The duration of illness affecting the children ranges from one day up to 120 days in all the three groups. However, most children reported the duration of illness for three days. Of the total children in each of the groups, the number of children who have fallen ill for three days in WSD+GM, WSD and No Input groups are 18 (16.2%), 16 (12.6%) and 7 (6.0%), respectively.

Treatment Seeking

In most cases of illness, the treatment is sought from private doctors in all the three groups. Other common place is *anganwadi* centre.

Medicines taken from government doctor/hospital, taking home-made medicines and taking medicines from medical store directly is comparatively less in all the three groups.

5

Household Socio-Economic Profile

The household socio-economic profile is generated on the basis of many indicators. Watershed development is the main input contributing to the current status of these features as observed in the households of children aged 0–5 years in the study.

Population

Population Composition

In total, 326 households of children aged 0–5 years have been covered under the study. This includes 100 households in WSD+GM group, 117 in WSD group and 109 in No Input group. In these households there are 2,113 members having 1,071 girls and 1,042 boys in the age group 0–5 years. Out of the total 326 households covered in the study, 315 (96.6%)

belong to Hindu religion and the remaining 11 (3.4%) are Muslim, Buddhist and Jain.

Community Composition

The communities are of two types: the tribal (settled and nomadic) and the caste. The caste communities are divided into three categories, viz., agriculturist caste, artisan caste and service caste, as per traditional Hindu caste structure.

The tribal communities include Thakar, Bhil, Mahadeo Koli, Lamani, Banjara, Pardhi, Phase Pardhi, Wadar, Bhamta, Kaikadi, Kolhati, Beldar and Gosavi.

The agriculturist caste includes the Maratha community.

The communities belonging to artisan caste are Sonar, Shimpi, Sutar, Mali, Lohar, Kumbhar, Teli, Kasar, Gurav and Gavli.

The communities from the service caste include Mahar, Mang, Matang, Chambhar and Bhangi.

The community composition and the population in the three studied groups are as follows:

Table 5.1

Community composition

Community	*Category*	*WSD+GM*	*WSD*	*No Input*
Agriculturist caste	Number	68	65	79
	Percentage	68.0	55.6	72.4
Tribal community	Number	16	28	16
	Percentage	16.0	23.9	14.7
Artisan caste	Number	12	19	10
	Percentage	12.0	16.2	9.2
Service caste	Number	4	5	4
	Percentage	4.0	4.3	3.7
Total	Number	100	117	109
	Percentage	100	100	100

Type of Family

There are 229 (70.2%) joint families and 97 (29.8%) are of nuclear type.

Women-Headed Households

In all, 8 (8.0%), 6 (5.1%) and 4 (3.7%) households in WSD+GM, WSD and No Input groups respectively have a woman as the head of the household, who are widows.

Marital Status

There are no child marriages in the area. Hence, the age group 1–10 years has been excluded in the analysis of the marital status. Adolescent marriages (age 14, 15 years) have been found and are common among the tribal and the Mahar communities.

Table 5.2

Marital status of the family members of the children

Marital status	*Category*	*WSD+GM*	*WSD*	*No Input*
Married	Number	390	434	383
	Percentage	86.9	86.3	88.2
Separated	Number	0	1	0
	Percentage	0.0	0.2	0.0
Widow/er	Number	19	18	12
	Percentage	4.2	3.6	2.8
Unmarried	Number	40	49	39
	Percentage	8.9	9.7	9.0
Missing	Number	0	1	0
	Percentage	0.0	0.2	0.0
Total	Number	449	503	434
	Percentage	100	100	100

There is not much difference in the marital status of the individuals across groups.

Primary Occupation

The primary occupation is considered for all the members aged 15 years onwards. Among all, 69.3 per cent of the individuals have agriculture as their main occupation, 12.5 per cent are engaged in labour work and 6.1 per cent in service. The remaining (12.1%) comprises students, unemployed, retired, housewives or doing any other business.

Landownership

There are more number of landless households in No Input group followed by WSD and WSD+GM groups.

Overall the households owning land in the range of less than and up to four acres are more in all the three groups.

Table 5.3

Landownership of the households of children aged 0–5 years

Landownership	*Category*	*WSD+GM*	*WSD*	*No Input*
Landless	Number	12	21	20
	Percentage	12.0	17.9	18.3
Small farmers ≤ 4 acres	Number	78	74	59
	Percentage	78.0	63.2	54.1
Middle level farmers >4 ≤ 8 acres	Number	7	14	19
	Percentage	7.0	12.0	17.4
Large farmers> 8 acres	Number	3	7	11
	Percentage	3.0	6.0	10.1
Missing	Number	0	1	0
	Percentage	0.0	0.9	0.0
Total	Number	100	117	109
	Percentage	100	100	100

More number of households in WSD+GM and WSD groups own land less than or equal to four acres. This indicates that the households in watershed villages have more number of small farmers.

There are more number of households owning land more than or equal to four acres but lesser than eight acres in No Input group. This indicates that the households in watershed villages have more number of small farmers.

Housing

Type of House

The houses found in the area are of different types as shown below in Table 5.4. The types are defined on the basis of houses found in the area. The houses made up of cement concrete are considered as of *pucca* type. The cement houses which have traditional roofing (tiled roofs) or flooring (mud flooring) are considered as of semi-*pucca* type while the mud houses (with no use of cement) are considered as of *kutcha* type. It is the use of cement that primarily defines the type of house here.

Table 5.4

Type of house

Groups	*Category*	*Type of house*			
		Pucca	*Semi-pucca*	*Kutcha/hut*	*Total*
WSD+GM	Number	67	14	19	100
	Percentage	67.0	14.0	19.0	100
WSD	Number	50	24	43	117
	Percentage	42.7	20.5	36.8	100
No Input	Number	69	19	21	109
	Percentage	63.3	17.4	19.3	100

The data indicates that more number of *pucca* houses are found in WSD+GM group, followed by No Input and WSD groups. More number of semi-*pucca* and *kutcha*/hut type of houses are found in WSD group, followed by No Input and WSD+GM groups.

Cattle Shed

The cattle shed is located either inside or outside the house. The following table presents the location of cattle shed:

Table 5.5

Location of cattle shed

Groups	*Category*	*Inside house*	*Outside house*	*Not applicable*	*Total households*
WSD+GM	Number	7	65	28	100
	Percentage	7.0	65.0	28.0	100
WSD	Number	5	69	43	117
	Percentage	4.3	59.0	36.8	100
No Input	Number	4	71	34	109
	Percentage	3.7	65.1	31.2	100

Thus, more houses belonging to WSD+GM and No Input groups have cattle shed outside the house and their number is almost equal.

Among the households having cattle shed outside house, more have closed type of cattle shed in WSD+GM group. The households having closed type of cattle shed are 39 (60.0%), 24 (34.8%) and 29 (40.8%) in WSD+GM, WSD and No Input groups, respectively.

More number of open type of cattle shed are found in WSD and No Input groups. The numbers of such households are 26 (40.0%), 45 (65.2%) and 42 (59.2%) in WSD+GM, WSD and No Input groups, respectively.

The closed type of cattle shed is fully covered by walls and has roof. The open type has no walls and may or may not have roof.

Kitchen

The following table indicates that more number of houses from WSD group have kitchen inside the house followed by the No Input and WSD+GM groups.

Table 5.6

Location of kitchen

Groups	*Category*	*Inside house*	*Outside house*	*Missing*	*Total households*
WSD+GM	Number	67	32	1	100
	Percentage	67.0	32.0	1.0	100
WSD	Number	90	27	0	117
	Percentage	76.9	23.1	0.0	100
No Input	Number	78	31	0	109
	Percentage	71.6	28.4	0.0	100

The households having kitchen inside house may or may not have separate room whereas the households having kitchen outside house may or may not have closed type of kitchen.

There are 25 (37.3%) households out of 67 in WSD+GM group, 31 (34.4%) households out of 90 in WSD group, 35 (44.9%) households out of the 78 in No Input group that have separate room for kitchen.

There are 19 (59.4%) households out of 32 in WSD+GM group, 17 (63.0%) households out of 27 in WSD group, 25 (80.6%) households out of 31 in No Input group that have closed type of kitchen.

Household Assets

There are 11 assets in total as shown in Table 5.7 below:

Table 5.7

Availability of household assets

Assets available	*Category*	*WSD+GM (N = 100)*	*WSD (N = 117)*	*No Input (N = 109)*
Electricity	Number	84	74	81
	Percentage	84.0	63.2	74.3
Cycle	Number	46	61	49
	Percentage	46.0	52.1	45.0

Cont'd...

...Cont'd

Motorcycle	Number	37	40	40
	Percentage	37.0	34.2	36.7
Motor pump	Number	43	48	60
	Percentage	43.0	41.0	55.0
Television	Number	54	55	58
	Percentage	54.0	47.0	53.2
Sewing machine	Number	14	10	12
	Percentage	14.0	8.5	11.0
Telephone (landline+mobile)	Number	25	32	25
	Percentage	25.0	27.4	22.9
Refrigerator	Number	6	3	8
	Percentage	6.0	2.6	7.3
Tractor	Number	0	1	4
	Percentage	0.0	0.9	3.7
CD/VCD/ DVD player	Number	14	6	18
	Percentage	14.0	5.1	16.5
Tempo/truck	Number	5	1	5
	Percentage	5.0	0.9	4.6

'N' indicates the total number of households.

A summary of Table 5.7 would help us rapidly grasp the data. The table is summarised by allotting scores to each asset available. For each of the assets, a score of 1 is given to that group wherein its ownership is highest.

Thus, the WSD+GM group scores 5, due to the availability of assets like electricity, motorcycle, television, sewing machine and tempo/truck with more number of households. Likewise, the WSD group scores 2 due to the availability of assets like cycle and telephone. And, the No Input group score 4 due to the availability of assets like motor pump, tractor, refrigerator and CD/VCD/DVD player.

The table further indicates that the WSD+GM and the No Input groups have similar status with respect to household assets than WSD group. The difference in the assets like motorcycle and tempo/truck between WSD+GM and No Input group is negligible (below 0.5%).

A higher availability of assets like motor pump and tractor in No Input group reveals better agriculture development in this group. A higher availability of tempo/truck with the WSD+GM group also indicates a better agriculture development in this group as these vehicles are mostly used for transportation of onion and tomatoes to the market.

Fuel

The following table indicates that more number of households from No Input group use LPG/natural gas, followed by WSD+GM group and WSD groups. There is not much difference between the No Input and WSD+GM groups. The WSD group is highly dependent on wood as fuel followed by WSD+GM and No Input groups. However, more than one fuel type are used by the households from all three groups.

Grass/agriculture waste is not much used as fuel except one household from WSD group.

Table 5.8

Use of fuel

Type of fuel used	*Category*	*WSD+GM (N = 100)*	*WSD (N = 117)*	*No Input (N = 109)*
LPG/natural gas	Number	19	5	22
	Percentage	19.0	4.3	20.2
Kerosene	Number	5	1	5
	Percentage	5.0	0.9	4.6
Wood	Number	86	113	87
	Percentage	86.0	96.6	79.8
Grass/agricultural waste	Number	0	1	0
	Percentage	0.0	0.9	0.0
Missing	Number	2	1	2
	Percentage	2.0	0.9	1.8

'N' indicates the total number of households.

Drinking Water

The status of drinking water of the households can be understood on the basis of availability and accessibility of drinking water, sources of drinking water and water purification followed by households.

Accessibility of Drinking Water

The following table shows that the number of households having access to drinking water in the campus itself is more in WSD+GM group, followed by No Input and WSD groups. There is not much difference between the WSD+GM and the No Input groups in this respect. The number of households travelling up to one kilometre for drinking water is more in WSD and No Input groups while the number of households from WSD+GM group is lesser in this case. Some households from watershed villages (WSD+GM and WSD) have to travel more than a kilometre to fetch water while there are few such households in No Input group.

Table 5.9

Distance covered to fetch drinking water

Groups	*Category*	*In the campus*	*0–1 km*	*>1 km*	*Total*
WSD+GM	Number	48	48	4	100
	Percentage	48.0	48.0	4.0	100
WSD	Number	48	63	6	117
	Percentage	41.0	53.8	5.1	100
No Input	Number	50	57	2	109
	Percentage	45.9	53.3	1.8	100

Purification of Drinking Water

The information regarding households that carry out purification of drinking water at any time of the year is presented in Table 5.10. While 'yes' indicates purification is done at any time of the year and 'no' indicates not doing at all.

Table 5.10
Households undertaking purification of drinking water

Groups	*Category*	*Yes*	*No*	*Total*
WSD+GM	Number	80	20	100
	Percentage	80.0	20.0	100
WSD	Number	56	61	117
	Percentage	47.9	52.1	100
No Input	Number	52	57	109
	Percentage	47.7	52.3	100

The table indicates that 80 per cent of the households from WSD+GM group undertake water purification any time in the year as against 48 per cent of the households in WSD and No Input groups. It is also observed that straining of the drinking water through cloth is the most common method used for water purification across all the three groups in all the three seasons of the year. The next most common method used is chlorination of water.

Water Shortages

The information relating to the households that have faced water shortages in one year preceding the survey is displayed in Table 5.11. Water shortages faced by a household is indicated with a mention of 'Yes' and if not by 'No'.

Table 5.11
Households facing water shortages

Groups	*Category*	*Yes*	*No*	*Total*
WSD+GM	Number	39	61	100
	Percentage	39.0	61.0	100
WSD	Number	52	65	117
	Percentage	44.4	55.6	100
No Input	Number	19	90	109
	Percentage	17.4	82.6	100

The table reveals that the numbers of households that have faced water shortage are more in WSD group followed by WSD+GM and much lesser in No Input group as compared to the two watershed villages. The overall water shortage in the study villages ranges from one to four months from March to June with severity in the months of April, May and part of June until the commencement of rain.

Sources of Drinking Water

The well water is the common source of drinking water in all the three groups under study. The other sources differ from group to group. In WSD+GM group, 50 (50.0%) households use well water for drinking followed by tap as the most common source with 34 (34.0%) households making use of it. In WSD group, 48 (41.0%) households use well water for drinking followed by handpump with 28 (23.9%) households making use of it. In No Input group, 58 (53.2%) households use well water for drinking followed by borewell with 24 (22.0%) households making use of it.

Borewell, lift irrigation, handpump, tanker, open sources of water like stream, percolation tanks, river/pond are the sources of water for 16 (16.0%), 41 (35.0%) and 27 (24.8%) of the households in WSD+GM, WSD and No Input groups respectively.

Out of the households reporting water scarcity in the past one year, private well is the main source of drinking water during the period of water shortage in WSD+GM and No Input groups. The number of these households is 14 (35.9%) and 8 (42.1%) respectively. In No Input group, tanker is the main source with 24 (46.2%) households depending on it.

Education

Table 5.12 below reveals the educational status of the households covered in the present study:

Table 5.12

Educational status

Groups	*Educational status*	*Category*	*Age groups (in years)*					
			5–10	*11–15*	*16–20*	*>20*	*Missing*	*Total*
WSD+GM	Illiterate	Number	36	0	1	136	0	173
		Percentage	20.8	0.0	0.6	78.6	0.0	100.0
	Literate	Number	27	1	1	20	0	49
		Percentage	55.2	2.0	2.0	40.8	0.0	100.0
	Primary	Number	19	10	2	53	1	85
		Percentage	22.3	11.8	2.4	62.3	1.2	100.0
	Secondary	Number	0	7	12	98	0	117
		Percentage	0.0	6.0	10.2	83.8	0.0	100.0
	Matric	Number	0	1	9	49	0	59
		Percentage	0.0	1.7	15.2	83.1	0.0	100.0
	Higher educated	Number	0	0	2	49	0	51
		Percentage	0.0	0.0	3.9	96.1	0.0	100.0
	Missing	Number	3	0	0	0	0	3
		Percentage	100.0	0.0	0.0	0.0	0.0	100.0
	Total	Number	85	19	27	405	1	537
		Percentage	15.8	3.5	5.0	75.5	0.2	100.0
WSD	Illiterate	Number	55	1	3	150	0	209
		Percentage	26.3	0.5	1.4	71.8	0.0	100.0
	Literate	Number	36	1	3	38	0	78
		Percentage	46.1	1.3	3.9	48.7	0.0	100.0
	Primary	Number	5	8	4	58	0	75
		Percentage	6.7	10.7	5.3	77.3	0.0	100.0
	Secondary	Number	0	6	10	69	1	86
		Percentage	0.0	7.0	11.6	80.2	1.2	100.0
	Matric	Number	0	1	11	78	0	90
		Percentage	0.0	1.1	12.2	86.7	0.0	100.0

Cont'd...

...Cont'd

	Higher educated	Number	0	1	11	51	0	63
		Percentage	0.0	1.6	17.4	81.0	0.0	100.0
	Missing	Number	0	17	0	0	0	17
		Percentage	0.0	100.0	0.0	0.0	0.0	100.0
	Total	Number	96	35	42	444	1	618
		Percentage	15.5	5.7	6.8	71.8	0.2	100.0
No Input	Illiterate	Number	31	1	3	115	1	151
		Percentage	20.4	0.7	2.0	76.2	0.7	100.0
	Literate	Number	33	2	2	33	0	70
		Percentage	47.1	2.9	2.9	47.1	0.0	100.0
	Primary	Number	13	7	4	44	0	68
		Percentage	19.1	10.3	5.9	64.7	0.0	100.0
	Secondary	Number	1	10	5	82	0	98
		Percentage	1.0	10.2	5.1	83.7	0.0	100.0
	Matric	Number	0	1	9	63	1	74
		Percentage	0.0	1.4	12.1	85.1	1.4	100.0
	Higher educated	Number	0	0	0	58	0	58
		Percentage	0.0	0.0	0.0	100.0	0.0	100.0
	Missing	Number	15	0	0	0	4	19
		Percentage	78.9	0.0	0.0	0.0	21.1	100.0
	Total	Number	93	21	23	395	6	538
		Percentage	17.3	3.9	4.3	73.4	1.1	100.0

Notes: *Literate:* It includes those persons who know to read and write or have done some schooling up to 1st, 2nd or 3rd standard.
Illiterate: It includes those persons who cannot read and write.
Primary education: This includes schooling up to standard four.
Secondary education: This includes schooling up to standard seven.
Matric: This includes schooling up to tenth standard.
Higher studies: This includes completion of standard twelve and above education.

The number of individuals in the age group 5–10 years seeking primary education are 19 (22.3%), 5 (6.7%) and 13 (19.1%) in WSD+GM, WSD and No Input groups respectively.

The number of individuals in the age group 11–15 years seeking secondary school education are 7 (6.0%), 6 (7.0%) and 10 (10.2%) in WSD+GM, WSD and No Input groups respectively. The number of individuals in the age group 16–20 years seeking matric level education are 9 (15.2%), 11 (12.2%) and 9 (12.1%) in WSD+GM, WSD and No Input groups respectively. The number of individuals who are more than 20 years of age and have done higher studies are 49 (96.1%), 51 (81.0%) and 58 (100.0%) in WSD+GM, WSD and No Input groups respectively.

Watershed Development

A question in the household schedule pertained to the benefits of watershed development as perceived by the watershed groups (WSD+GM and WSD) of villages and the manner in which they have benefited. The following table gives the information across the watershed villages:

Table 5.13

Perceived benefits of watershed development

Benefits	*Category*	*WSD+GM*	*WSD*
Land treatment	Number	45	51
	Percentage	45.0	43.6
Horticulture/ floriculture	Number	11	3
	Percentage	11.0	2.6
Social development activity	Number	21	3
	Percentage	21.0	2.6
Labour	Number	48	70
	Percentage	48.0	59.8

Notes: Land treatment: This includes minor water and soil conservation treatments in the farms.

Social development activity: This includes participation in self help group and their activities.

The total number of households in WSD+GM and WSD groups is 100 and 117 respectively. The households that perceive no benefits from watershed development are 28 (28.0%) and 30 (25.6%) in WSD+GM and WSD groups respectively.

The table indicates that most of the households obtained benefit through employment as labourers in carrying out work for watershed development. The households from WSD group have benefited the most. The other important benefit obtained was from land treatment. About 44–45 per cent in both the groups have benefited from land treatment. Some have benefited from social development activity, which is a part of watershed development, for example, facilitation and initiation of SHGs. Among these, the number of households from WSD+GM group is more than the WSD group. Few have also got the benefits in the form of development of horticulture and floriculture. Among these, the number of households from WSD+GM group is more than the WSD group.

The number of households that perceive getting more than one benefit from watershed development are 35 (35.0%) and 36 (30.8%) in WSD+GM and WSD groups respectively. The number of households that have got all the mentioned benefits from watershed development are only 5 (5.0%) in WSD+GM group. There are no such households in WSD group.

Agriculture

The number of households having agricultural land are 88 (88.0%), 96 (82.1%) and 89 (81.7%) in WSD+GM, WSD and No Input groups respectively.

Type of Agriculture

The following table shows that agriculture is of rainfed type in watershed development groups (WSD+GM and WSD) as compared to No Input group. More number of households from No Input group possesses more irrigated land than watershed development groups. Overall, the households with partially irrigated lands are more.

Table 5.14

Type of agriculture

Type of agriculture	*Category*	*WSD+GM*	*WSD*	*No Input*
Rainfed	Number	47	46	30
	Percentage	53.4	47.9	33.7
Partially irrigated	Number	35	42	49
	Percentage	39.8	43.8	55.1
Fully irrigated	Number	6	8	8
	Percentage	6.8	8.3	9.0
Missing	Number	0	0	2
	Percentage	0.0	0.0	2.2
Total	Number	88	96	89
	Percentage	100.0	100.0	100.0

Area of Crops

As the following table indicates, in No Input group, the area covered under crops is more than the watershed development groups. The farm area of the middle level farmers from WSD group is slightly more under crop cultivation than the other two groups.

Table 5.15

Area covered under crops

Total area under crops	*Category*	*WSD+GM*	*WSD*	*No Input*
≤ 4 acres	Number	27	38	53
	Percentage	30.7	39.6	59.5
> 4 ≤ 8 acres	Number	12	19	16
	Percentage	13.6	19.8	18.0
> 8 acres	Number	5	7	11
	Percentage	5.7	7.3	12.4
Missing	Number	44	32	9
	Percentage	50.0	33.3	10.1
Total	Number	88	96	89
	Percentage	100	100	100

Crops and their Production

Bajra/pearl millet is the staple crop. The other important crops grown in the area are groundnut, sorghum, onion, tomato, potato, cotton, pulses like red gram *(tur)*, green gram *(mug)*, moth beans *(mat)*, flower and fruit crops.

In WSD+GM group, 20 (20.0%) households are growing groundnut crop. The total production is 284 quintals (qtl). In WSD group, 6 (5.1%) households are growing groundnut crop. Total production is 156 qtl. In No Input group, 4 (3.7%) households are growing groundnut crop. The total production is 11 qtl.

In WSD+GM group, 26 (26.0%) households are growing onion crop. The total production is 1,439 qtl. In WSD group, 31 (26.5%) households are growing onion crop. The total production is 5,210 qtl. In No Input group, 27 (24.8%) households are growing onion crop. The total production is 2,207 qtl.

In WSD+GM group, 6 (6.0%) households are growing tomato crop. The total production is 197 qtl. In WSD group, 8 (6.8%) households are growing tomato crop. The total production is 1,507 qtl. In No Input group, 4 (3.7%) households are growing tomato crop. The total production is 55 qtl.

Thus, the production of groundnut crop is comparatively more in WSD+GM group. The production of onion and tomato is higher in WSD group although the number of households growing these crops is near to WSD+GM and No Input groups.

As mentioned by the respondent villagers and also observed during the survey, there has been an increase in the yield after watershed development, especially in groundnut, onion and tomato.

Income from Sale of Agriculture Yield

Overall, the income from agriculture yield is in the range of ₹ 20,000 to 50,000 per annum in all the three groups. The maximum income exceeds ₹ 1 lakh in WSD+GM and WSD groups whereas in No Input group the maximum income is up to ₹ 1 lakh.

Horticulture and Floriculture

In WSD+GM villages, 88 households own land whereas in WSD and No Input villages there are 96 and 89 households respectively having land.

In comparison, there are more number of households from WSD+GM group, that is, 16 (18.2%), which practise horticulture. In WSD and No Input groups of villages, 8 (8.3%) and 3 (3.4%) of the households, respectively, are engaged in horticulture. Sweet lime, orange and pomegranate are the fruits usually grown in the orchards in the area.

More number of households from No Input group, that is, 16 (18.0%), are engaged in floriculture as compared to WSD+GM and WSD groups. In WSD+GM group, 8 (9.1%) and in WSD group 1 (10.0%) are found to be engaged in floriculture. *Bijli,* Marigold *(Zendu)*, *Aster* and *Shewanti* are the flowers usually grown in the field.

The average annual income from horticulture ranges from ₹ 1,000 to more than ₹ 1 lakh while from floriculture it is up to ₹ 50,000.

Kitchen Garden

The status with regard to kitchen garden is almost similar in WSD+GM and No Input groups. The number of households from WSD+GM and No Input groups having kitchen garden are 14 (14.0%) and 17 (15.6%) respectively. Kitchen garden is not so popular in WSD group with only 9 (7.7%) households having it. Some vegetable crops like brinjal, garlic, curry leaves, coriander, colocasia leaves (*Alu*), lemon, chillies, pumpkin and fruit crops like mango, chickoo, guava, papaya, coconut are grown around the households.

It is also observed that although watershed development activity has taken place in the two groups (WSD+GM and WSD), the kitchen garden growers are quite less.

Livestock

More number of households in WSD+GM and WSD groups own livestock as compared to No Input group. In WSD+GM and WSD groups the number of households owning livestock are 76 (76.0%) and 85 (72.6%) respectively whereas in No Input group there are 74 (67.9%) households owning livestock.

The animal ownership in WSD+GM group is comparatively more. The following table gives the status of livestock ownership across studied groups:

Table 5.16

Individual livestock ownership

Livestock ownership	*Category*	*WSD+GM N = 76*	*WSD N = 85*	*No Input N = 74*
Bull	Number	48	45	37
	Percentage	62.3	52.9	50.0
Local cow	Number	38	43	26
	Percentage	50.0	50.6	35.1
Crossbred cow	Number	27	43	40
	Percentage	35.5	50.6	54.1
Buffalo	Number	12	12	4
	Percentage	15.8	14.1	5.4
Goat	Number	54	51	41
	Percentage	71.1	60.0	55.4
Hen	Number	25	19	16
	Percentage	32.9	22.4	21.6

'N' indicates the livestock owning households.

More households in WSD+GM group own bulls, buffaloes, goats and hens in comparison to WSD and No Input groups. The households having local cows are almost similar in WSD+GM and WSD groups. The number of households having crossbred cows is more in No Input group, followed by WSD group.

The number of bulls possessed by the households ranges from one to four. More number of households own two bulls in all the three groups. Of the households owning bulls, 36

(75.0%), 36 (80.0%), 31 (83.8%) households in WSD+GM, WSD and No Input groups respectively own two bulls.

The number of crossbred cows owned ranges from 1 to 10. Out of the households having crossbred cows in WSD+GM and No Input groups, more number of the households have two cows. In all, 10 (37%) and 18 (45.0%) households in these groups have two cows respectively. In WSD group, more number of households (44.2%) have one cow.

Of the households owning crossbred cows, 14 (51.9%) in WSD+GM group and 17 (39.5%) in WSD group started domesticating them only after watershed development.

Use of Livestock

Most households (95%) in all the three groups make use of bulls in their own farms. The number of households that make use of local cows for household milk consumption in WSD+GM, WSD and No Input groups are 35 (92.1%), 38 (88.4%) and 20 (76.9%), respectively.

The number of households that make use of crossbred cows for household milk consumption in WSD+GM, WSD and No Input groups are 22 (81.5%), 36 (83.7%) and 29 (72.5%) respectively. The number of households using crossbred cows for commercial use that is for selling milk in WSD+GM, WSD and No Input groups are 3 (11.1%), 5 (11.6%) and 11(27.5%), respectively.

The number of households using buffaloes for household milk consumption in WSD+GM, WSD and No Input groups are 10 (83.3%), 7 (58.3%) and 4 (100.0%), respectively.

The goats and hens in all the three groups have more of domestic uses. Goat milk is used for household milk consumption and hens are primarily used in rituals and the eggs for household consumption.

Purchase of Livestock

The data on the livestock purchased by the households are collected for a period of one year preceding the survey. Out of the total households, that is, 100, 117 and 109 in WSD+GM, WSD and No Input groups respectively, more number of households in WSD+GM, i.e., 30 (30%), purchased livestock in the last one year preceding the survey as compared to WSD and No Input groups, i.e., 14 (12.0%) and 20 (18.3%), respectively.

Animals purchased by the households in the studied villages during the last year include bulls, local cows, crossbred cows, buffaloes and goats. The purchase of bulls is more in WSD group as compared to WSD+GM and No Input groups. A total of 6 (42.9%) households in the WSD group purchased bulls as compared to 12 (40.0%) and 5 (25.0%) of the households in WSD+GM and No Input groups, respectively.

The purchase of crossbred cows is more in No Input groups. Households that purchased crossbred cows in No Input group are 15 (75.0%) as compared to 7 (50.0%) and 12 (40.0%) of the households in WSD and WSD+GM groups, respectively.

More households (16.7%) in WSD+GM group purchased buffaloes compared to No Input groups (5.0%). Households in WSD group did not purchase any buffalo.

The number of local cows and goats purchased are very low as compared to other animals.

Reasons for purchasing livestock are depicted below in Table 5.17:

Table 5.17

Reasons for purchasing livestock

Reason	*Category*	*WSD+GM*	*WSD*	*No Input*
Commercial	Number	2	1	3
	Percentage	6.7	7.1	15.0
Dairy business	Number	0	5	10
	Percentage	0.0	35.7	50.0
Farming	Number	10	5	2
	Percentage	33.3	35.7	10.0
Farming and Milk production	Number	0	1	2
	Percentage	0.0	7.1	10.0

Cont'd...

...Cont'd

Milk production	Number	18	2	3
	Percentage	60.0	14.3	15.0
Total	Number	30	14	20
	Percentage	100	100	100

Notes: Commercial: The term includes activities like rearing of young livestock and selling and/or giving bulls for ploughing on rent.
Dairy business: The term includes business of selling milk to dairies.
Farming: This includes use of bulls in own fields.
Milk production: This includes milk production of cattle and goat for domestic consumption.

More number of households in No Input group, that is, 3 (15.0%), have purchased livestock for commercial purpose as compared to the other groups, that is, 1 (7.1%) and 2 (6.7%) households in WSD and WSD+GM group, respectively. The percentage for WSD+GM and WSD group is approximately the same.

A total of 10 (50.0%) of the households from No Input group have purchased animals for dairy business as compared to 5 (35.7%) households from WSD group. None of the WSD+GM households have purchased animals for dairy business.

However, households purchasing animals for farming are more in WSD+GM and WSD groups (33.3% and 35.7% respectively) as compared to No Input group (10.0%).

There are more number of households, that is, 18 (60%) in WSD+GM group who purchased animals for milk production as compared to WSD and No Input groups. In WSD and No Input groups their number is 2 (14.3%) and 3 (15.0%), respectively.

In WSD+GM group, while more households have purchased animals for milk production, in WSD the main reason for purchase was dairy as well as farming. However in No Input group, more number of households have purchased animals for dairy business only.

Milk Production

Local Cow

The milk production ranges from less than 1–5 litres per day in all the three groups for local cows. However, more number of households in all the three groups take less than one litre of milk production per day.

Crossbred Cow

The milk production of crossbred cows varies from less than 1 litre to more than 50 litres per day across the three groups. However, there are more number of households who have per day milk production in the range of 5–10 litres. The number of households in this range are 5 (17.9%), 9 (20.5%) and 8 (18.6%) in WSD+GM, WSD and No Input groups, respectively.

Buffalo

Among the households owning buffaloes, more number of them are taking 1–5 litres of milk production: 4 (33.3%) in WSD + GM, 4 (33.3%) in WSD and 1 (25%) in No Input group.

Milk Selling

Among the households owning crossbred cows, 21 (77.8%), 34 (79.1%) and 28 (70.0%) households in WSD+GM, WSD and No Input groups, respectively, sell milk. The quantity of milk sold ranges from less than 1 litre to 50 litres per day. However, more number of households in all the three groups sell milk in the range of 5–10 litres per day. The places where the milk is sold are shown in Table 5.18.

Thus, more number of households in No Input group sell milk in milk cooperatives. Their number in WSD+GM, WSD and No Input groups are 10 (47.6%), 20 (58.8%) and 19 (67.9%), respectively.

The market rate for milk varies from ₹ 7–14 per litre.

Table 5.18
Places of milk selling

Reason	*Category*	*WSD+GM*	*WSD*	*No Input*
Own dairy	Number	10	13	8
	Percentage	47.6	38.3	28.6
Milk cooperative	Number	10	20	19
	Percentage	47.6	58.8	67.8
Missing	Number	1	1	1
	Percentage	4.8	2.9	3.6
Total	Number	21	34	28
	Percentage	100	100	100

Period of Milk Selling

Some households have started selling milk recently whereas some households are selling for more than 20 years. In WSD+GM group more number of households, i.e., 8 (38.1%) have been selling milk since 5 to 10 years. In WSD group, more numbers of households, i.e., 18 (52.9%) have been selling milk since 2 to 5 years. In No Input group more number of households, i.e., 11 (39.3%) have been selling milk since 5 to 10 years.

There are 2 (7.1%) households in No Input group who are selling milk since more than 20 years whereas the number of such households in WSD+GM and WSD groups is 1 (4.8%) and 1 (2.9%), respectively.

Income from Selling Milk

The income from selling milk ranges from less than one thousand rupees to more than one lakh rupees per annum in all three groups.

In WSD+GM group, more number of households, i.e., 5 (23.8%) earn annual income in the range of ₹ 20,001–50,000, whereas WSD and No Input groups, more number of households, i.e., 13 (38.2%) and 10 (35.7%) get annual income in the range of ₹ 10,001–20,000.

In WSD+GM and WSD groups there are 2 (9.5%) and 2 (5.9%) households each who are earning in the range of ₹ 50,001–1,00,000 whereas in No Input group there are no such households. Similarly, there are 2 (9.5%) households in WSD+GM group with income more than one lakh whereas there are no such households in WSD and No Input groups.

Livestock Selling

The respondents were also asked about the sale of livestock in the households one year preceding the survey and its purpose. Out of all the households in WSD+GM group, 13 (13.0%) have sold livestock in one year where the corresponding number of such households in WSD and No Input groups was 7 (6.0%) and 19 (17.4%) respectively.

The purpose of selling livestock is given in the following table:

Table 5.19

Purpose of selling livestock

Purpose	*Category*	*WSD+GM*	*WSD*	*No Input*
Distress sale	Number	7	6	8
	Percentage	53.8	85.7	42.1
Business	Number	1	0	10
	Percentage	7.7	0.0	52.6
Social	Number	2	0	1
	Percentage	15.4	0.0	5.3
Missing	Number	3	1	0
	Percentage	23.1	14.3	0.0
Total	Number	13	7	19
	Percentage	100	100	100

There are more number of households in WSD+GM and WSD groups who have sold their livestock during distress than those from the No Input group. In No Input group more number of households have sold livestock for the sake of business. There is one household in WSD+GM group who has sold livestock for business but there is no such household in WSD group.

Income from Livestock Selling

The average income from livestock selling is up to ₹ 50,000 among all three groups. In WSD+GM and WSD groups more number of households have earned income up to ₹ 5,000. In No Input group more number of households have had income in the range of Rs.10,001– 20,000.

Poultry

The number of households having poultry is 24 (24.0%), 28 (23.9%) and 33 (30.3%) in WSD+GM, WSD and No Input groups, respectively.

More number of households in the three groups rear poultry for domestic consumption than for any other reason. The number of households rearing poultry for domestic consumption is 22 (91.7%), 22 (78.6%) and 32 (97.0%) in WSD+GM, WSD and No Input groups, respectively. The other reasons for poultry rearing are social purpose, business and selling.

More number of households from WSD group, that is, 6 (21.4%) than those from WSD+GM, that is, 2 (8.3%) and No Input, that is, 2 (6.1%) rear poultry for business purpose.

Migration

The number of households migrating out of village at any time in the year is 22 (22.0%) in WSD+GM, 28 (23.9%) in WSD and 17 (15.6%) in No Input groups. Thus, migration in No Input group is comparatively lesser.

Reasons for Migration

As the following table shows more number of households in all the three groups migrate for labour work. Migration for labour work is comparatively more in No Input group.

There are few households who migrated for service work in WSD+GM group. Few households from WSD group have migrated for business purpose.

Table 5.20

Reasons for migration

Purpose	*Category*	*WSD+GM*	*WSD*	*No Input*
Labour	Number	17	26	16
(agriculture/other)	Percentage	77.3	92.9	94.1
Business	Number	1	2	1
	Percentage	4.5	7.1	5.9
Service	Number	4	0	0
	Percentage	18.2	0.0	0.0
Total	Number	22	28	17
	Percentage	100	100	100

Distance of Migration

The distance of migration is mostly in the range of 101 to 500 km. In WSD+GM and No Input groups the number of such households are 9 (40.9%) and 11 (64.7%), respectively. In WSD group 3 (10.7%) households migrate in this range whereas more households, that is, 13 (46.4%), migrate in the range of 50 km.

Period of Migration

The period of migration is mostly up to six months. The number of households migrating for this period in the three groups is 5 (22.7%), 5 (17.9%) and 5 (29.4%) in WSD+GM, WSD and No Input groups, respectively.

Food Shortages

The data on food shortages was collected by an inquiry of the situation one year preceding the survey with the following details:

There are 7 (7.0%), 7 (6.0%) and 7 (6.4%) households in WSD+GM, WSD and No Input groups respectively who reported eating 'less than usual' in a year preceding the survey. The reported period of shortage is in the range of 1 to 60 days.

A common reason stated by the households from all the three groups for eating less than usual were lack of work during the rainy season and therefore inadequate money to purchase grain or food.

There are 2 (2.0%), 4 (3.4%) and 4 (3.7%) households in WSD+GM, WSD and No Input groups respectively who reported eating 'no food at all' on some days in the year preceding the survey.

The period of having 'no food at all' ranged from 1 to 15 days. In WSD+GM group, one household each reported not eating at all for 3 and 6 days in past one year. In WSD group two households each reported not eating at all for 2 and 4 days in past one year. In No Input Group one household each reported not eating at all for 1, 2, 5 and 15 days.

Loans

Loan Amount

In WSD+GM group 51(51.0%) households, in WSD group 43 (36.8%) households and in No Input group 47 (43.1%) households have taken loans. The amount of loan taken ranges from ₹ 5,000 to 2 lakh and more for all the households taken together.

More number of households in all the three groups have taken loans in the range of ₹ 20,001–50,000. There are 13 (25.5%) households from WSD+GM group, 15 (34.9%) from WSD and 16 (34.0%) households from No Input group who have taken loans ranging from ₹ 20,001 to 50,000. Two (4.7%) households of WSD group have taken loan of more than ₹ 2 lakh.

Reasons for Taking Loans

The reasons for taking loan across the three groups include medical, wedding, housing, agriculture and business.

In WSD+GM group more number of households, i.e., 11 (21.6%) have taken loans for housing purpose. After housing, the most common reason stated in this group is agriculture with 17.6 per cent households.

In WSD group more number of households, i.e., 11 (25.6%) have taken loans for agriculture purpose. After agriculture, the most common reason stated in this group is housing with 20.9 per cent households.

In No Input group more number of households, i.e., 32 (68.1%) have taken loans for agriculture purpose. After agriculture, the other reasons stated in this group are housing and business with 8.5 per cent households each.

Sources of Loan

As the data in the following table reveal among all the three groups, more number of households in No Input group have taken loans from institutions. Their numbers are 34 (66.6%), 29 (67.4%) and 40 (85.1%) in WSD+GM, WSD and No Input groups, respectively.

Table 5.21

Sources of loan

Sources of loan	*Category*	*WSD+GM* *N = 51*	*WSD* *N = 43*	*No Input* *N = 47*
Institution	Number	34	29	40
	Percentage	66.7	67.4	85.1
Moneylender	Number	2	8	4
	Percentage	3.9	18.6	8.5
Friend/relative	Number	16	9	3
	Percentage	31.4	20.9	6.4
Missing	Number	1	0	0
	Percentage	2.0	0.0	0.0

Among all the three groups, more number of households from WSD group take loans from moneylenders. The number of such households in WSD+GM, WSD and No Input groups are 2 (3.9%), 8 (18.6%) and 4 (8.5%), respectively.

The households taking loans from friends and relatives are more in WSD+GM and WSD groups as compared to No Input group.

Ways of Loan Repayment

More number of households have repaid the loans in the form of cash in all the three groups. Other form of repayment is doing labour work on farms.

Thus, this chapter has widely discussed the household socio-economic data of children aged 0–5 years covered under the study.

6

Profile of Mothers of Children Aged 0–5 Years

This chapter presents the data on the socio-economic background of the mothers of children aged 0–5 years covered under the study. The total number of mothers are 126. Their numbers in WSD+GM, WSD and No Input groups are 99, 118 and 109, respectively.

Age of Mother

The age group of more number of mothers in all three groups under study is 21–25 years. Mothers who are less than 20 years of age are 9 (9.1%), 12 (10.2%) and 8 (7.3%) in WSD+GM, WSD and No Input groups, respectively.

Education

There are more mothers who have completed their schooling up to secondary level (class seven). The number of such mothers is more in WSD+GM group. The distribution is as follows: 44

(44.4%), 34 (28.8%) and 38 (34.9%) in WSD+GM, WSD and No Input groups, respectively.

Occupation

Agriculture is the main occupation of most of the mothers of children aged 0–5 years in all the three groups. The mothers reporting agriculture as their occupation in WSD+GM, WSD and No Input groups are 57 (57.6%), 65 (55.1%) and 68 (62.4%), respectively. Other occupations are service, labour work, dairy and business.

Institutional Membership

More number of mothers from WSD+GM group have institutional membership. For example, as the table below shows, the numbers of mothers having membership of SHGs across study groups are 31 (31.3%), 14 (11.9%) and 20 (18.3%) in WSD+GM, WSD and No Input groups, respectively.

Table 6.1

Individual institutional membership

Institution	*Category*	*WSD+GM (N = 99)*	*WSD (N = 118)*	*No Input (N = 109)*
SHGs	Number	31	14	20
	Percentage	31.3	11.9	18.3
Village Watershed Committee	Number	1	0	0
	Percentage	1.0	0.0	0.0
Gram Panchayat	Number	0	0	0
	Percentage	0.0	0.0	0.0
Dairy Cooperative	Number	1	0	0
	Percentage	1.0	0.0	0.0
Any other committee	Number	1	0	0
	Percentage	1.0	0.0	0.0
None	Number	68	104	89
	Percentage	68.7	88.1	81.7

More numbers of mothers are members of SHGs. The number of such members is more in WSD+GM group, followed by No Input and WSD groups.

Besides SHGs, the women from WSD+GM group only are members of other institutions like village watershed committee, Gram Panchayat, Dairy Cooperative Society and other committee.

Entrepreneurship

The following table indicates that dairy is the most common source of entrepreneurship than other activities.

Table 6.2

Entrepreneurial activities

Institution	*Category*	*WSD+GM (N = 99)*	*WSD (N = 118)*	*No Input (N = 109)*
Poultry farming	Number	6	4	3
	Percentage	6.1	3.4	2.8
Sheep rearing	Number	0	1	0
	Percentage	0.0	0.8	0.0
Dairy	Number	20	29	14
	Percentage	20.2	24.6	12.8
Sale of cash crops	Number	0	0	7
	Percentage	8.1	0.0	6.4
Other	Number	2	1	1
	Percentage	2.0	0.8	0.9
None	Number	65	83	86
	Percentage	65.7	70.3	78.9

More mothers in WSD group are found to be involved in dairy as entrepreneurial activity, followed by WSD+GM and No Input groups. Similarly, more mothers from WSD+GM group are found to be involved in sale of cash crops, followed by No Input and WSD groups.

Decision-Making

The number of mothers involved in taking the decision about the kind of crops to be grown in the field are mainly from watershed groups, viz., 52 (61.9%) and 60 (61.2%) from WSD+GM and WSD groups respectively and 37 (48.1%) from No Input group. Here, the total number of mothers was 84, 98 and 77 in WSD+GM, WSD and No Input groups, respectively.

Exposure to Mass Media

As the following table shows more number of mothers from WSD+GM group read newspapers and watch television followed by No Input and WSD groups. There is a wide variety of newspapers and magazines read by the mothers in the villages. They are related to general news, political issues, health, women, festival magazines, spiritualism etc.

Table 6.3

Exposure to mass media

Media type	*Category*	*WSD+GM (N = 99)*	*WSD (N = 118)*	*No input (N = 109)*
Reading newspaper at least once a week	Number	49	44	47
	Percentage	49.5%	37.3%	43.1%
Reading magazines	Number	17	7	5
	Percentage	17.2%	5.9%	4.6%
TV watching	Number	60	57	54
	Percentage	60.6%	48.3%	49.5%

Television Programmes Watched by Mothers

More mothers in all the three groups reported watching entertainment programmes most. The number of mothers watching entertainment programmes on television is more among WSD and No Input groups. It is 42 (70.0%), 44 (77.2%) and 42 (77.8%) in WSD+GM, WSD and No Input groups, respectively.

The number of mothers watching news is more in WSD+GM group. It is 15 (25.0%), 5 (8.9%) and 2 (3.7%) in WSD+GM, WSD and No Input groups, respectively.

Age at Marriage, Cohabitation and First Childbirth

Most of mothers in all the three age groups reported 'less than 18 years' as the age of marriage. Mothers reporting 'less than 18 years' as the age of marriage are 80 (80.8%), 92 (78.0%) and 85 (78.0%) in WSD+GM, WSD and No Input groups, respectively. The age of first cohabitation is the same as age at marriage for most of the mothers in all the three groups and it is 'less than 18 years'.

Most of the mothers in all the three age groups have reported their age at first childbirth in the age 18–21 years. Their number are 49 (49.5%), 55 (46.6%) and 56 (51.4%) in WSD+GM, WSD and No Input groups respectively.

Pregnancy

It is observed that full completion of nine months of pregnancy period reduces with the number of births of children in all the groups. The number of mothers not completing the period is comparatively more in WSD+GM group as observed from the births of second child onwards up to the birth of fourth child.

Table 6.4

Completion of pregnancy periods of mothers

Child	*Category*	*WSD+GM (N = 99)*	*WSD (N = 118)*	*No Input (N = 109)*
Child 1	Number	96	114	101
	Percentage	97.0	96.6	92.7
Child 2	Number	72	94	79
	Percentage	72.7	79.7	72.5
Child 3	Number	17	30	27
	Percentage	17.2	25.4	24.8
Child 4	Number	5	9	4
	Percentage	5.1	7.6	3.7
Child 5	Number	2	1	1
	Percentage	2.0	0.8	0.9

Mothers Reporting Child Deaths

The number of mothers reporting child deaths are 6 (6.1%), 5 (4.2%) and 4 (3.7%) in WSD+GM, WSD and No Input groups, respectively.

Some of the causes of child deaths as reported by the mothers are hepatitis, premature deliveries, accident, bleeding from nose and malaria.

Of the 16 child deaths reported by mothers from all the three groups, 10 (62.5%) are girls and 6 (37.5%) are boys. Out of the 10 girl child deaths, three are from WSD+GM group, five from WSD group, and two from No Input group.

7

Village Infrastructure

Every village has a single Gram Panchayat. The villages – Karjule Pathar and Gunjalwadi – are two separate revenue villages, have two separate *anganwadis* but fall under one Gram Panchayat. Hence, although the WSD+GM group consists of four villages, the village profile of one village is clubbed with another village from WSD group. The village profile information for the two villages remains the same. Hence, in the analysis and presentation of findings of the village infrastructure, the WSD+GM group is considered as having three villages and WSD and No Input groups have four villages each.

Distance of Villages from Taluka and District Headquarters

The average distance of villages included in WSD+GM, WSD and No Input groups from taluka is 34, 18 and 26 km, respectively and from district headquarters is 51, 69 and 61 km, respectively.

It indicates that villages from WSD+GM group among all are comparatively far from taluka place but closer to district headquarters, while the villages from WSD group are far from district place and closer to taluka place. However, the No Input group villages are far from both the taluka and the district places.

Approach Roads to Taluka

The villages from WSD and No Input groups are connected to the taluka place through a *pucca* road (developed tar roads). In WSD+GM group, the villages are connected to the taluka place partly through *pucca* road and partly through semi-*pucca* (not fully developed) road while a *pucca* road connects villages of all groups to the district headquarters. It indicates that overall the roads connecting to taluka level and district places are well developed.

The total population of WSD+GM, WSD and No Input groups is 3,791, 8,672 and 5,947, respectively.

Sex Ratio

The sex ratio of population from WSD+GM, WSD and No Input groups is 974, 981 and 958 respectively.

The sex ratio of children of age 0–5 years from WSD+GM, WSD and No Input groups is 783, 878 and 948, respectively.

There is a skewed sex ratio of children aged 0–5 years in all groups. There is a declining trend in watershed groups, viz., WSD+GM and WSD.

Literacy Rate

The average literacy rate in villages from WSD+GM, WSD and No Input groups is 76, 60 and 63 per cent, respectively.

Communication

Half of the villages from WSD and No Input groups have post office facility available within the village whereas in WSD+GM group, only 2 (66.7%) of the villages have such facility.

Likewise, 2 (66.7%), 1 (25.0%) and 3 (75.0%) of the villages in WSD+GM, WSD and No Input groups respectively have public telephone facility available within the village.

All the villages in all the three studied groups have private telephone facility available within the village.

Two villages from WSD+GM group have computer facility available. There is no internet facility in these villages.

Transportation

There are 2 (66.7%), 2 (50.0%) and 3 (75.0%) villages of WSD+GM, WSD and No Input groups respectively having transportation facility available from village to taluka place, while there are 1 (33.3%), 1 (25.0%) and 2 (50.0%) villages of WSD+GM, WSD and No Input groups respectively having transportation facility available from village to district place. The frequency of these services is more than once daily.

Education

All the villages in all the three groups under study have schooling facility up to 4th standard available within the village.

There are 3 (100%), 2 (50%) and 2 (50%) villages in WSD+GM, WSD and No Input groups respectively that have schooling facility upto 7th standard available within the village. In WSD villages where this facility is not available within the village, children have to travel average distance of 3.5 km for going to the nearest secondary school whereas in No Input group, children have to travel little less, i.e., 2.5 km, for the same.

There are 2 (66.7%) and 1 (25.0%) villages of WSD+GM and WSD groups respectively having high school (8–10 standard) facility available within the village. In No Input village where high school is not available within the village, children have to travel 5 km for going to the nearest high school. Even in WSD+GM and WSD groups, children have to travel average distance of 2.8 and 2.5 km, respectively, for going to the nearest high school.

None of the villages in all the three groups have degree college facility available within the village. In WSD+GM villages, students have to travel average distance of 19.3 km for going to the nearest degree college while from WSD and No Input groups, the travelling distance is 15.5 and 17.3 km respectively for the same.

Anganwadi

All the villages in all the three groups under study have *anganwadi* centres within the village.

Health Care Facility

Primary Health Care

The average distance of Primary Health Centre (PHC) from WSD+GM, WSD and No Input group villages is 25.7, 18.5 and 12 km, respectively.

The average distance of PHC sub-centre from WSD+GM, WSD and No Input group villages is 14.3, 4 and 3 km, respectively.

Private Doctors

There is only one private doctor (MBBS) each in villages from WSD+GM (33.3%) and No Input (25.0%) groups available within the village. In WSD group, no such facility is available.

There are 1 (33.3%), 1 (25.0%) and 3 (75.0%) villages from WSD+GM, WSD and No Input groups respectively having private doctor (BAMS/BUMS/BHMS) available within the village.

There are 1 (33.3%) and 2 (50.0%) villages of WSD+GM and No Input groups respectively having trained TBA (Traditional Birth Attendant) in the village. Again, in WSD group, this facility is not available.

There are 2 (66.7%) and 1 (25.0%) villages of WSD+GM and WSD groups respectively having traditional medical practitioners in the village. This facility is not available in villages of No Input group.

Medical Shop

There is no medical shop in the villages from WSD+GM group while there are 1 (25%) and 2 (50%) villages from WSD and No Input groups, respectively that have a medical shop in the village.

Land Utilisation

The total village area of WSD+GM, WSD and No Input groups is 3,113.5, 4,952.3 and 4,685.4 hectares, respectively.

The total cultivable area in villages from WSD+GM, WSD and No Input groups is 1,823, 3,888 and 3,781 hectares respectively. The total forest area of WSD+GM, WSD and No Input groups is 842.3, 589.2 and 328 hectares, respectively. The total barren land area of WSD+GM, WSD and No Input groups is 296.1, 283.9 and 167.7 hectares, respectively.

In many villages barren land and/or forest land is used as pasture land. Pasture lands are not separately demarcated.

Total Area of Watershed

The total area covered under watershed development in WSD+GM and WSD groups is 3,250 and 6,173 hectares respectively.

Table 7.1

Total area of watershed and population

Comparative group	*Total area of watershed*	*Total population*
WSD+GM	3250 ha	3791
WSD	6173 ha	8672

Animal Care Services

The average distance of nearest government veterinary clinic from WSD+GM, WSD and No Input villages is 12.3 km, 4 km and 6.8 km, respectively.

The average distance of nearest private veterinary clinic from WSD+GM, WSD and No Input villages is 12.3 km, 4 km and 4 km, respectively (in WSD and No Input groups, data of 1 and 2 villages respectively is missing).

While 1 (33.3%), 3 (75.0%) and 3 (75.0%) villages of WSD+GM, WSD and No Input groups respectively have government artificial insemination facilities available in the village (or the service provider can be called whenever the need arises), 1 (33.3%), 3 (75.0%) and 2 (50.0%) villages of WSD+GM, WSD and No Input groups respectively have private/NGOs artificial insemination facilities available in the village (or the service provider can be called whenever the need arises).

Market

None of the villages in all the three groups have daily market in the village. None of the villages in WSD+GM and No Input groups have weekly market in the village. Only 1 (25%) village in WSD group has weekly market in the village.

The average distance of the nearest place from the market for WSD+GM, WSD and No Input group villages is 8, 7.3 and 5 km respectively.

Public Distribution System

All the villages in WSD+GM and No Input groups have a ration shop while in WSD group, 3 (75%) villages have ration shop.

Hotel

There are 1 (33.3%), 2 (50.0%) and 1 (25.0%) villages of WSD+GM, WSD and No Input groups respectively having a *tapri* (a tea stall where some snack items are also available) within the village. There are 2 (66.7%), 3 (75.0%) and 3 (75.0%) villages from WSD+GM, WSD and No Input groups respectively having a small hotel in the village. One village (25%) each from WSD and No Input groups has a big hotel in the village.

Milk Dairy

There are four dairies each in villages from WSD+GM and WSD groups and six in No Input group.

Local Governance

All the villages from WSD+GM and No Input group fall under single Gram Panchayat. In WSD group, there are two villages that have single and two have group Gram Panchayat.

The average number of GP members in WSD+GM, WSD and No Input groups is 7, 9 and 7 respectively. While the average number of males in WSD+GM, WSD and No Input groups is 5, 6 and 6 respectively, the average number of females in WSD+GM, WSD and No Input groups is 2, 3 and 2 respectively in Gram Panchayat.

Bank and Credit

None of the villages in WSD+GM and WSD group have bank facility available within the village. 1(25%) village in No Input group has bank available in the village. The average distance of the nearest place where bank is available is 8 km, 7 km and 5.3 km for WSD+GM, WSD and No Input groups respectively (in No Input group, this is applicable for villages where bank is not available).

Two villages each in WSD+GM (66.7%) and No Input (25.0%) groups have credit society available within the village. There is no credit society in the villages from WSD group.

Self Help Groups (SHGs)

The total number of women SHGs in WSD+GM, WSD and No Input groups is 28, 34 and 25, respectively.

The average number of women members in SHG from WSD+GM, WSD and No Input groups is 14, 9 and 6 respectively.

The total number of men SHGs in WSD+GM, WSD and No Input groups is 4, 1 and 0 respectively.

The activities of women's SHGs include cooking of *anganwadi* food, loan distribution to members and maintaining its records.

The women SHG members from WSD+GM group also run a private agency of provision of gas cylinders. They also undertake the work of distribution of ration cards, insurance for women, run a small-scale business of grinding spices (chilly powder and masala) and public call office (PCO/telephone booth). Two women members are health volunteers and undertake growth monitoring of children aged 0–5 years.

In WSD group, a few tribal women SHG members have started business of goat rearing. In No Input group, the women SHG members from one village run tuition classes for children up to 10th standard.

Other NGOs Working in Studied Villages

In 1 (33.3%) and 2 (50.0%) of the villages of WSD+GM and No Input groups, NGOs are working. In watershed villages the number of NGOs excludes WOTR.

In WSD+GM village, namely, Bhoyre Pathar, NGO named Pruthvi Sanstha has been working since 2003. The work of this organisation includes teaching children in age group 0–13 years about value-based education *(sanskar)*, exercises, yoga and art and craft (*hasta kala*).

In No Input villages two NGOs, namely, 'Chaitainya Gramin Mahila Yuvak va Balvikas Sanstha' and 'Don Bosco Vikas Kendra' are working in Astagaon and Walwane village since year 2008 and 2005 respectively. The work of NGO in

Astagaon is formation of and guidance to SHGs. The work of NGO in Walwane includes taking tuition classes for children from 1–10th standard and teach them about body cleanliness.

Other Village Level Institutions

None of the villages in all the three comparative groups have village level institutions for storage and food security like grain bank, seed bank or fodder bank.

8

Development Profile and Child Nutrition

A Comparative Perspective

Profile of Child Nutrition

There is not much difference across groups under study in terms of height-for-age. More than 60 per cent of the children in all the three groups are normal for their height. However, the number of children from No Input group is slightly more than the others. The difference between the No Input group and the WSD+GM group is not much. In terms of weight-for-age, the WSD group has more number of normal children (54%) than other groups, while there is not much difference in WSD+GM and No Input groups in this regard.

Protein is an important component of diet that is responsible for the increase in height while consumption of carbohydrates and fats in the diet adds up to weight. The results indicate that protein is comparatively available in the diet but there is a need to increase energy giving food to children aged

0–5 years. If we look at the results of actual energy and protein consumption of these children and compare with the Recommended Dietary Allowance by ICMR-NIN, we find that there is a consistency in the results obtained from the two methods, viz., anthropometry and diet survey. The protein consumption of children from all the groups is better than energy consumption. Secondly, the actual protein and energy consumption of children from No Input group is better than the WSD+GM and WSD groups. Within the watershed groups, it is the WSD+GM group where the actual protein and energy consumption of children is better than the WSD group as exemplified by the results of diet survey. Although these are the results of diet survey, the same is not getting reflected in terms of weight.

In terms of weight-for-age, the WSD group shows a better status than the other two groups. The food frequency results have indicated that the consumption of eggs, legumes and fruits is more in WSD group than other groups. This is also the group where there is more number of children who are completely immunised for their age. An important finding is eating less than usual among children has not been reported in WSD group.

However, the weight-for-age status is still not satisfactory in all the three groups. There is a need to explore as to why there is not much weight gain in children in the villages while the growth of height shows a better status.

The average birth weight of children from all the three groups is around 2.6 kg. As birth weight of 2.5 kg and above is considered as normal, the results indicate that the children aged 0–5 years had a normal weight at the time of their birth and the status is similar in all the comparative groups. However, the birth weight needs to be increased to have more number of healthy babies.

Immunisation coverage is better in WSD group and is comparatively lower in WSD+GM group. According to National Family Health Survey (NFHS-3) reports, only 43.5 per cent of children in the age group 12–23 months are fully vaccinated whereas the figure for the state of Maharashtra is 58.8 per cent as a whole and 49.8 per cent for rural Maharashtra. 5 per cent of children in rural Maharashtra have not received any single vaccines (IIPS, 2006).

One of the reasons is the location of *anganwadi* where immunisation takes place. One village usually has one *anganwadi*. The *anganwadi* is not closer to all hamlets of the village. Some hamlets are located at a large distance from the *anganwadi*. The houses are on the hilly terrain and are isolated from the main village. In such situations, the children from such hamlets do not attend *anganwadi*. The distance could be one of the reasons due to which the parents/care givers are not able to bring their children for immunisation. In fact the *anganwadi* workers take great efforts to reach out to these children and provide immunisation to the child if he/she misses out. Observations and informal discussions with people indicate there is greater awareness among the people regarding immunisation and see to it that the child gets immunised against diseases. However, the status needs further improvement to achieve full coverage. Migration of the parents for employment could be another reason when they are likely to miss on the immunisation schedule of the children.

The morbidity results indicate that the children do not get ill very often. Illnesses like cough and cold, fever are common. No major illnesses have been reported among the children age 0–5 years in all the three groups. This contributes to better nutritional and health status of the children. The number of children who have fallen ill at least once since birth are much lesser in WSD group but more in No Input group. Probably morbidity could be one of the factors because of which the children from the No Input and WSD+GM group have lost weight while the status of WSD group is better in terms of weight-for-age.

Regarding child deaths, there are more number of girl child deaths in watershed development groups as reported by the mothers. At the same time, the sex ratio of watershed groups shows a declining trend as compared to the No Input group. It would not be right to comment on the reasons with the limited data that is available in the study. However, the study highlights a correlation in the two phenomena. There is a need to understand the causes of the girl child deaths and the declining sex ratio in watershed villages.

The results of food insufficiency faced by children indicates that there are more children in WSD+GM group and few in No Input group while the WSD group has no such children. The food insufficiency faced by children indicates the severity of food shortage in the households. Even, while the socio-economic status of WSD+GM and No Input groups is better than WSD group, there are children facing food insufficiency for quite a long period. The controversy needs to be looked into carefully and the reasons have to be explored. The data on food shortage (in terms of eating less than usual) faced by the households also shows that the WSD+GM group has slightly more number of such households.

Development Profile of the Studied Groups: A Summary

A brief analysis of the profile of the three groups is presented here for two main reasons. One is to have a quick grasp of the results so as to discuss the differences. Second reason is to understand the correlations within the data, specifically the child nutrition and the development profile across the groups.

WSD+GM Group

The WSD+GM group is a group of small farmers. More number of households have not only *pucca* type of houses, they have also closed type of cattle shed.

Electricity, motorcycle, television and sewing machine are available in more number of houses than other groups. However, CD/VCD/DVD player and refrigerator are found in some houses. More number of households have tempo/truck than other groups. In terms of assets, the two groups, viz., WSD+GM and No Input are almost similar.

There is not much difference between the No Input and the WSD+GM group regarding the ownership of LPG/natural gas.

More households than those from other groups have access to drinking water within their premises. More households than those from other groups undertake purification of water at any time of the year. This is the group to which tap water is available besides well water while other groups make use of handpumps and borewells. During the period of water shortage, tanker dependency is comparatively lesser than other groups.

More number of children are seeking primary education and matriculation than other groups. The WSD+GM group has benefited from labour work, land treatment, socio-development activity and from horticulture and floriculture after watershed development. The group has received multiple benefits unlike the WSD group.

There are more households with agriculture of rainfed type. The area covered under crops is lesser than the other two groups. The number of households growing groundnut crop are more and its production is more than the other groups. The agricultural income exceeds ₹ 1 lakh. More households are engaged in horticulture. Households having kitchen garden are more than that from WSD group but less than No Input group.

More number of households in WSD+GM group own bulls, local cows, buffaloes, goats and hens. Household milk consumption is more than other groups. More households have purchased buffaloes in one year than other groups. The purchase of livestock here is for farming and for household milk production. Households have sold livestock in one year during distress.

The number of households migrating out for employment is comparatively lesser than WSD group but more than No Input group. More households from WSD+GM group than others have migrated out for service besides labour work.

Households that had food less than usual in one year are few and the difference is not much in all three groups. Households that did not have food at all are less than No Input group but slightly more than WSD group.

More households have taken loan. The purpose of taking loan is for housing and agriculture. Most of the loan has been taken from friends and relatives.

There are more mothers who have completed their school education up to secondary level (class seven). Agriculture is their prime occupation. More number of mothers from WSD+GM group have institutional membership, i.e., they are members of Self-Help Groups (SHGs). More mothers from WSD+GM group are involved in poultry farming and sale of cash crops. More mothers are involved in taking the decision about the kind of crops to be grown in the field. More number of women from WSD+GM group read newspapers, magazines and watch television. The number of mothers watching news is more in WSD+GM group. The number of mothers not completing full-term pregnancy (gestation) period is comparatively more in WSD+GM group as observed from the births of second child onwards up to the birth of fourth child. More mothers have reported child death in this group.

The distance of the villages from taluka place is comparatively more but lesser from district place. The roads connecting to taluka place are not so well developed.

The total population of the villages is comparatively less. The sex ratio of children aged 0–5 years is lesser than all other groups.

The average literacy rate is better than other groups. The availability of telephone facility is more than that of WSD group but lesser than that of No Input group.

Education facility is available at a comparatively greater distance than other groups.

The Primary Health Centre and the sub-centre are located far from the villages.

There are few private doctors and trained TBAs in the villages. The number is lesser than No Input group. There is no medical shop in the villages. Out of the village area, the total cultivable area is lesser than all other groups (68.8%). The total area covered under watershed development is lesser than that of WSD group.

Government and private veterinary clinics are far off. Artificial insemination facility is available to few villages. None of the villages have a daily or weekly market and the market is far off.

Ration shop is available in all the villages. There is no big hotel as such but small hotels are there in the villages.

The watershed villages have equal number of milk dairies.

None of the villages have bank facility available within the village. Credit society is available in half of the villages. The number of women SHGs is more than that of No Input group but less than WSD group. The women SHG are involved in lot of activities.

One of the villages has an NGO other than those of WOTR and Sampada Trust.

WSD Group

The WSD group is a group of small and middle level farmers.

More number of households have semi-*pucca and kutcha*/hut type of houses. More have open type of cattle shed.

Assets like cycle and telephone are possessed by more number of households.

The WSD group is highly dependent on wood as fuel.

More households have to travel up to 1 km and more to fetch drinking water. About half of the households undertake purification of water all round the year. In this respect the two groups, viz., WSD and No Input are similar. The group has faced more water shortage in one year than other groups. They make use of wells and hand pumps as sources of drinking water.

WSD group is lagging behind in education from primary to higher studies.

The households from WSD group have benefited from labour work the most and land treatments due to watershed development.

Agriculture is of rainfed nature than No Input group but lesser than WSD+GM group. Irrigation is comparatively more than WSD+GM group but little less than No Input group. The farm area of the middle level farmers from WSD group is slightly more under crop cultivation than the other two groups. The number of households growing onion and tomato crops are more and its production is more than that of other groups. The agricultural income exceeds ₹ 1 lakh. The households practicing horticulture is more than those of No input group. Kitchen garden is not popular.

The number of local cows in WSD group is almost equal to that of WSD+GM group. More households have one cow than other groups. The households having two bulls are more than WSD+GM group. Milk selling is similar in watershed villages. The purchase of bulls is more in WSD group as compared to WSD+GM and No Input groups. The purchase of crossbred cows is more in WSD group than WSD+GM group but lesser than No Input group. The purchase of livestock here is for farming and for dairy business. Milk selling is more in this group than others.

The number of households migrating out for employment is comparatively more. Most do so for labour work. Few households have migrated out for business work.

Households that had food less than usual in one year are few and the difference is not much in all three groups. Households that did not have food at all are comparatively less.

The households that have taken loan are less than WSD+GM group. The purpose of taking loan is for agriculture and housing. Most of the loan has been taken from money lenders.

Few mothers have completed their school education up to secondary level (class seven). Agriculture is their prime occupation. Few mothers have institutional membership, that is are members of Self-Help Groups (SHGs). More mothers are involved in taking the decision about the kind of crops to be grown in the field and is equivalent to WSD+GM group. There are less number of women having exposure to mass media. The number of mothers watching entertainment programmes is more in the group and the status is similar to that of No Input group. The number of mothers not completing the full gestation period is comparatively more in WSD+GM group as observed from the births of second child onwards up to the birth of fourth child. More mothers have reported child death.

The distance of the villages from taluka place is comparatively lesser than all others. The roads connecting to taluka and district places are well developed. The villages are far from district place.

The total population of the villages is comparatively more. The sex ratio of children aged 0–5 years is less.

The average literacy rate is lesser than other groups. The availability of telephone facility is comparatively less. Transportation is well developed. Education facility is available nearby.

The Primary Health Centre and the sub-centre are comparatively nearer to the villages. Private doctors and trained TBAs are comparatively lesser in the villages. One of the villages has a medical shop. Government and private veterinary clinics are close by. Artificial insemination facility is available to most villages.

Out of the village area, the total cultivable area is lesser than the No Input group but more than WSD+GM group (78.5%). The total area covered under watershed development is more.

One of the villages is a place for weekly market. Ration shop is not available in one of the villages.

One small and one big hotel each is available in the villages.

The watershed villages have equal number of milk dairies.

None of the villages have bank facility available within the village. There is no credit society in the villages.

There are more number of women SHGs in the village. The tribal women SHG members have started business of goat rearing and have completed a year.

No Input Group

The No Input group is a group of large farmers. More number of houses are of *pucca* type. More number of houses have separate room for kitchen.

More number of households possess tractor, motor pumps and also tempo/truck. More number of houses have LPG/natural gas, refrigerator, CD/VCD/DVD player.

Water is available within the campus and some have to travel up to maximum 1 km to fetch water. Less than half of the people undertake purification of water at any time of the year. The group does not face water shortage much. However, during the period of water shortage, the tanker dependent households are more than WSD+GM group. For drinking, water from wells and borewells is used.

The number of students who have undertaken higher studies is more.

More number of households possess irrigated land. The area covered under crops is more than other groups. Some are engaged in floriculture. Kitchen garden plantation is comparatively more.

More number of households own crossbred cows and bulls. Sale of milk is more. The purchase of crossbred cows in one year is more in the group. More number of households have purchased animals for dairy business. More number of them sell milk in milk cooperatives. There are households selling milk more than 20 years. More number of them have sold livestock for business purpose in one year than others. More number of households among all possess poultry and mainly rear them for domestic consumption.

The number of households migrating out for employment is comparatively less. Those who migrate out do so for the purpose of labour work.

Some households face food shortage. Some have reported eating no food at all in a year and the period is comparatively more than other groups.

More number of households have taken loans for agriculture purpose. More number of households have taken loans from institutions like banks, cooperatives, SHGs, etc.

The number of mothers involved in agriculture is more. The number of mothers who have completed full gestation period (at least complete nine months) is less.

The sex ratio of children aged 0–5 years is comparatively better than watershed groups.

The villages connecting to taluka and district headquarters have pucca roads. More number of villages have public telephone facility available within the village. More villages have well developed transportation facility. Secondary school and high school facility is nearby. Primary health care is available at closer distance. More number of the villages have private doctors in the village. There are no traditional medical practitioners in the village. More number of villages have medical shop in the village. The total cultivable area out of the village area is more than other groups (80.7%). Private veterinary services are available close to the villages. Market is available at a near distance than other groups. One of the villages has a big hotel. There are more number of milk dairies in the villages. One of the villages has a bank in the village itself. Other villages have the facility available nearby the villages. In No Input group, the women SHG members from one village run tuition classes for children up to 10th standard. More number of villages have NGOs working in the village.

The nutritional status of children aged 0–5 years in the study region shows that more than 60 per cent of the children are normal for their height-for-age. Height-for-age is the impact indicator of watershed development. The results indicate that there is not much difference in the status across the comparative groups. However, the No Input group is slightly ahead of the watershed groups, viz., WSD+GM and WSD, that is, it has slightly more number of children with normal height-for-age. Within the watershed villages, WSD+GM group is slightly ahead of WSD group in terms of the number of children with normal height-for-age. The difference is not very critical.

Variations in Development and Child Nutrition among the Groups

The results indicate that the child nutritional status is better in the No Input group so far as height-for-age index is concerned. The factors that have contributed to better nutritional status as revealed by the study are as follows:

1. Agricultural development is much better than the watershed development groups, viz., WSD+GM and WSD groups. This is because the households from the No Input group possess more land than other groups. There are more number of large farmers. The land that they own is well irrigated. Rainfed land is less. Floriculture is well developed. The total cultivable area of the village is much more than the other groups. Because of more agricultural work, they have assets like tractor, tempo/truck and motor pump. More women are engaged in agricultural work than other groups. Due to availability of agricultural work, the number of members migrating out for employment is comparatively less.
2. Livestock development is also better than the other groups. There are more number of dairies. The number of crossbred cows are more. More households are engaged in dairy business. The sale and purchase of livestock is more for the purpose of business rather than distress.
3. Although it is a group where integrated watershed development has not taken place, minor water and soil conservation work have been undertaken by the government departments. This has been a limitation of the study. We did not find villages in Ahmednagar, which could be called as the real No Input, where no kind of water and soil conservation has taken place. Ahmednagar is a drought-prone district and after a severe drought in the year 1972, the government has been keen in water and soil conservation in the district. Besides there are irrigation development programmes. A survey of 20 villages was carried out to find out a village where no water and soil conservation input is given but the efforts were in vain.
4. There are situations where certain villages get the benefit of watershed development carried out in nearby places. For example, one of the No Input villages, namely, Jachakwadi is close to a village called Mhaswandi, where integrated watershed development has taken place and receives all the benefits. Jachakwadi is located on the plains while Mhaswandi is situated on the hilltop. The benefits of watershed development in Mhaswandi flow down to

Jachakwadi following the natural flow of water from top to bottom due to geographical reasons.

5. The households have a better socio-economic status as indicated by their housing, kitchen, cattle shed, asset possession.
6. The No Input villages have a well-developed infrastructure as indicated by the village profile.
7. The people have the capacity to take institutional loan.

These are some of the factors that have contributed to the overall development of the group and thereby to child nutrition. However there are certain weaknesses in the nutritional status of the children of this group. While more children are normal as per height-for-age, more children are underweight. This is a paradoxical situation. The results of mother's profile indicate that more mothers from this group than others are involved in agriculture. Probably due to heavy work load in agriculture for women, child feeding is getting neglected. The children lack in energy giving food.

An important finding related to food shortage is that food shortage exists although minimal in this group even when agricultural and livestock development is good.

A question that occurs is whether excessive selling of food crops can lead to the situation?

The WSD+GM group is closer to the No Input group. There is not much difference (0.9%) in the nutritional status of children as per height-for-age in the two groups.

The number of normal children from WSD group is comparatively less. The growth monitoring inputs are probably making the difference in the WSD+GM group as against the WSD group. However, the situation is reverse with respect to the nutritional status based on weight-for-age. In this case, the WSD group is faring well as compared to the other two groups. This is a paradox and the study cannot explain the reason for the situation based on the data collected. A further research is needed for a deeper understanding that will explain the paradoxical situation.

The results however reveal that the integrated watershed development is not making much difference in the status of child nutrition as well as the overall development of the watershed groups, viz., WSD+GM and WSD in comparison to the No Input group. The factors that are likely to be responsible for the situation are as follows:

1. The WSD+GM group and the WSD group are the groups of small and middle level farmers respectively. Land ownership is not much in these groups as compared to the No Input group.
2. The total cultivable area in the villages of these groups is lesser than that of the No Input group.
3. Watershed development can only create some buffer for soil moisture and groundwater. If the rainfall is not adequate, the benefits are not much in terms of development of irrigation potential. There is no data related to rainfall distribution in the villages of the different groups.
4. There is no data related to the type of land and soil in the different groups. The properties of land and soil too determine the extent of impacts of watershed development.
5. In WSD+GM group, the socio-economic status of the households is better as indicated by housing, cattle shed, kitchen, household and agricultural assets and in terms of children seeking education.
6. In terms of education, the status of WSD+GM group and the No Input group is almost similar. While WSD+GM group has more number of children seeking matric education than the No Input group, there are more children from No Input group who are involved in higher studies. The two groups also have a better socio-economic status than the WSD group, which could be one of the factors that has contributed to seeking higher education in these two groups. However the number of children seeking secondary education is lesser in all the three groups and more so in the WSD+GM group. Probably there are more drop-outs at the secondary level. There is a need to explore the reasons of school drop-out in watershed villages having better socio-economic status.
7. The watershed development seems to have impacted upon women in WSD+GM group. As a part of

socio-development activity certain changes are observed in women. For example, the differences in the mothers of children in WSD+GM setting it apart from the other groups. More mothers from WSD+GM group have institutional membership, are involved in decision-making and entrepreneurial activities, and have more exposure to mass media. Some of the negative changes are the declining sex ratio of children aged 0–5 years in watershed groups. There is a skewed sex ratio in the whole study region. However, the change is pronounced and faster in watershed villages. The completion of full-term pregnancy period reduces with the number of pregnancies. Also, there are more girl child deaths in WSD group.

8. The results reveal that there are large differences in the socio-economic statuses of the watershed groups, viz., WSD+GM and the WSD groups. The possible reasons are as follows:

   During the field visits, it was observed that in most villages from WSD group, the water and soil conservation structures were not functioning and needed repair and maintenance. In one of the villages, a mismanagement had occurred in the funds of the village watershed committee (VWC), in another village there were conflicts among the people, the people were not organised. This affects the functioning of VWC and other local institutions. This also affects people's participation in development activities, thereby causing negative effects/changes.

The selection of the villages for growth monitoring input is based on women's organisation. The villages where women are organised, that is SHGs are well developed and functioning are selected for child growth inputs. Usually, the village where people's organisations are functioning well are also the better developed watershed villages.

9

Watershed Development and Health

Watershed development will definitely lead to better health status of the people. In order to achieve this, the inputs of health should be consciously integrated with the programme of watershed development. The output of the programme should be measured from the perspective of health status of the community.

If health is considered as a goal of development, then, the community health should be priority in watershed development sector. This can be achieved if community health considerations are a part at all levels of the programme: planning, implementation and during measuring outputs.

The study was undertaken with the hypothesis stated as: "Integrated watershed development, strengthened with participatory child growth inputs, leads to reduction in the number of malnourished children aged zero to five years in the rural semi-arid region."

The following are some of the viewpoints that we would like to express with regard to the above mentioned hypothesis. The viewpoints are based on the study conducted and are open for criticism and discussion.

1. If we use the anthropometric index of height-for-age, which is the impact indicator of watershed development, the hypothesis is valid when the comparison takes place between the watershed groups, that is one where there is watershed development with an additional input of child growth monitoring (WSD+GM group) compared with the population where there is an input of watershed development but no additional input of growth monitoring (WSD group). When compared with the No Input group, the hypothesis does not stand. The No Input group/control group shows slightly better status (please refer to Table 4.1 for the overall nutritional status of the three studied groups with respect to height-for-age). The reasons are many. Some of the reasons that have come up from study are related to the socio-economic conditions in the control group (No Input). The conditions of higher land ownership, higher irrigation and higher socio-economic status, good village infrastructure of the No Input group are not comparable with the watershed groups, viz., WSD+GM and WSD groups. The study highlights that for comparison the comparative groups should have similar land related factors as mentioned. The similarities in socio-economic status and that of village profile are also important.
2. Since weight-for-age is not an impact indicator, the results based on this index do not matter much. However the paradoxical situation that has occurred by using an additional index of weight-for-age needs further exploration. The reasons need to be explored. Please refer to Table 4.1 where we find the overall nutritional status of three studied groups. Here, the WSD group despite not having additional input of growth monitoring shows better nutritional status as compared to other two groups.
3. One of the difficulties in the selection of No Input villages was that in Ahmednagar, there is hardly a village which can be called as not having any input related to water and

soil conservation. Villages with some or the other small structures built up for water and soil conservation are found everywhere and they do contribute a lot to the overall development of the village. This may have contributed to in part towards better development and nutritional status of the children from No Input group.

Appendices

Status of castes/tribes in Maharashtra

Community	*Status*	*Community*	*Status*
Maratha	Open	Chambhar	SC
Thakar	ST	Bhil	ST
Lamani/Banjara	DNT	Mahadeo Koli, Malhar Koli	ST
Teli	OBC	Koli	SBC
Kumbhar	OBC	Koshti	SBC
Sutar	OBC	Gowari	SBC
Lohar	OBC	Pardhi	ST, DNT
Gavli	OBC	Phanse Pardhi	ST, DNT
Gurav	OBC	Wadar	DNT
Mahar	SC	Sonar	OBC
Mang	SC	Shimpi	OBC
Matang	SC	Kolhati	NT
Kasar	OBC	Beldar	NT
Bhavsar	OBC	Gosavi	NT
Mali	OBC		

List of Tribes and Castes as Found in the Study Region

List of tribes

Tribal communities	
Settled tribe	*Nomadic tribe*
Thakar	
Bhil	
Mahadeo Koli	
	Lamani
	Banjara
	Pardhi
	Phase Pardhi
	Wadar
	Bhamta
	Kaikadi
	Kolhati
	Beldar
	Gosavi

List of castes as per traditional Hindu categorisation

Caste communities		
Agriculturist caste/ higher rung	*Artisan castes/ lower rung*	*Labourer caste/ Lowest rung*
Maratha		
	Sonar	
	Shimpi	
	Sutar	
	Mali	
	Lohar	
	Kumbhar	
	Teli	

Cont'd...

...Cont'd

	Kasar	
	Gurav	
	Gavli	
		Mahar
		Mang
		Matang
		Chambhar
		Bhangi

Status of communities in the Constitution

Community	*Status*	*Community*	*Status*
Maratha	Open	Chambhar	SC
Thakar	ST	Bhil	ST
Lamani/Banjara	DNT	Mahadeo Koli	ST
Teli	OBC	Koli	SBC
Kumbhar	OBC	Koshti	SBC
Sutar	OBC	Gowari	SBC
Lohar	OBC	Pardhi	ST, DNT
Gavli	OBC	Phanse Pardhi	ST, DNT
Gurav	OBC	Wadar	DNT
Mahar	SC	Sonar	OBC
Mang	SC	Shimpi	OBC
Matang	SC	Kolhati	NT
Kasar	OBC	Beldar	NT
Bhavsar	OBC	Gosavi	NT
Mali	OBC		

Source: Government of Maharashtra. Letter dated 25 May 2006. Declared official list of communities.

Interview Schedules

Introduction

The interview schedules have been prepared by reviewing NFHS-3 survey schedules, *Park's textbook of Preventive and Social Medicine*, discussions with Research Advisors, Clinician (Consultant) and from field experience of researchers and other WOTR officials.

Schedule for Children: 0–5 Years

I. Child's Personal Characteristics

Name of the child:

1. Sex: Boy ☐ Girl ☐
2. Age:
3. Birth order:
4. Birth weight (in kg.):

II. Food Intake

1. Feeding

i. When was the baby breast fed after delivery?

a. Within 2 hours ☐

b. Between 2–4 hours ☐

c. After 4 hours ☐

ii. Was the colostrum fed to the baby?

a. Yes ☐ b. No. ☐

Reason for yes: Reason for no:

2. What feeds were given to the baby after birth? Specify external items fed:

3. In case external item is fed, what is the reason for the same?

2. Supplementary food

i. Have you stopped breast feeding?

a. Yes ☐ b. No ☐

ii. If yes, at what age did you stop breast feeding?

iii. Do you give supplementary food to your child?

a. Yes ☐ b. No ☐

iv. If yes, ask: At what age of the child did you start giving supplementary food?

v. What is the supplementary food that you are giving? Make list.

vi. What food item did you start with?

vii. Is there any specific reason?

a. Yes ☐ b. No ☐

viii. What is the reason?

ix. If no, at what age would you start giving supplementary food?

x. Is there any specific reason?

a. Yes ☐ b. No ☐

xi. What is the reason?

3. Complementary feeding

i. Does the child consume milk other than breast feed?

a. Yes ☐ b. No ☐

ii. What type of milk is given?

a. Cow milk ☐ b. Buffalo milk ☐

c. Goat milk ☐ d. Milk powders ☐

e. Infant formula ☐ Brand name:

f. Packed milk ☐ Brand name:

iii. How is this milk fed to the child?

a. Bottle ☐ b. Container and spoon ☐

iv. If cow/goat/buffalo milk is given, is water added?

a. Yes ☐ b. No ☐

v. If water is added, ask about the proportion of milk to water:

Please tell about the Milk to water proportion

S. No.	External milk	Milk to water ratio			
		1:2	1:1	1:0.5	1:0.25
1.	Cow milk				
2.	Buffalo milk				
3.	Goat milk				
4.	Milk powder				
5.	Infant formula				
6.	Packed milk				
7.	Other:				

4. Previous day feeding schedule

S. No.	Quarter	Breast feed	Outside milk	Quantity of outside milk
1.	Morning			
2.	Mid morning			
3.	Lunch			
4.	Mid afternoon			
5.	Evening			
6.	Dinner			
7.	Bed time			
8.	Other:			

5. One day diet record: Recall method (Ask for the day before the interview)

Consumption of semi-solid/solid/liquid food

S. No.	Quarter	Food item	Ingredients of food	Quantity (gm/ml)	Place	
					At home	Outside home
1.	Morning					
2.	Mid Morning					
3.	Lunch					

Cont'd...

...Cont'd

4.	Mid Afternoon					
5.	Evening					
6.	Dinner					
7.	Bed time					
8.	Other					

Note: The quantities of food items, both solid and liquid should be measured with the help of measuring containers.

6. Food Frequency

S. No.	*Food item*	*Number of times in the following*		
		Day	*Week*	*Month*
1.	Fruits			
2.	Green Leafy Vegetables			
3.	Vegetables			
4.	Legumes			
5.	Pulses			
6.	Roots and Tubers			
7.	Porridge			
8.	Yogurts:			
9.	Baby food made by Farex, Cerelac, etc.			
10.	Egg			
11.	Fish			
12.	Chicken			
13.	Meat			

7. Food insufficiency in the past one year

i. Did the child face food shortage in the last one year?

Yes ☐ No ☐

If yes, please furnish the following information:

S. No.	*Months*	*Number of days of food shortage*	*Cause*
1.			
2.			
3.			

III. Health Information

1. Tonics

i. Do you give any tonic to the child?

a. Yes ☐ b. No ☐

ii. What tonic is given? (List out all the tonics or vitamin drops given to the child since birth).

iii. Who suggested you these tonics/vitamin drops?

iv. Do you give gripe water to the child?

a. Yes ☐ b. No ☐

v. If yes, what kind of product is given?

a. Home made ☐ b. Purchased ☐

If purchased, write brand name:

vi. How many times in a day?

vii. What is the purpose of giving gripe water?

2. Morbidity history of the child

S. No.	*Illness episodes*	*Age*	*Perceived cause*	*Duration*	*Treatment*	*Food*	*Feeds*
1.							
2.							
3.							
4.							
5.							

3. Status of Immunisation:

Name of the child:

S. No.	*Vaccine names*	*Status*
1.	BCG	
2.	POLIO 0	
3.	POLIO 1	
4.	POLIO 2	
5.	POLIO 3	
6.	DPT 1	
7.	DPT 2	
8.	DPT 3	
9.	MEASLES	
10.	VITAMIN A	

Mother's Schedule

Part A

I. Personal Characteristics

1. Name of the mother :
2. Age :
3. Education :
4. Occupation :

II. Institutional Membership

1. Are you a member of the following institution:

   i. Self help group ☐

   ii. Village watershed committee ☐

   iii. Gram panchayat ☐

   iv. Dairy cooperative ☐

   v. Other (specify):

   vi. None ☐

III. Entrepreneurial Activities

1. Are you involved in the following enterprises:

   i. Poultry farming ☐ ii. Sheep rearing ☐

   iii. Dairy ☐ iv. Sale of cash crops ☐

   v. Other: vi. None ☐

IV. Decision Making

1. Are you involved in taking decision about the kind of crops to be grown on your land?

   i. Yes ☐ ii. No ☐ iii. NA ☐

V. Exposure to Mass Media

1. How often do you read News paper?

   i. Daily ☐ ii. Once/twice a week ☐

   iii. Thrice a week ☐ iv. Never ☐

2. Name the newspapers you read
3. Do you read magazines?

   i. Yes ☐ ii. No ☐

4. Name the magazines you read:
5. Do you watch Television?

   i. Yes ☐ ii. No ☐

6. What programmes do you usually see?

   i. News ☐ ii. Educational ☐

   iii. Entertainment ☐

Part B
(Field Investigators to Note: Please Interview the Mother Alone)

VI. Demographic Information

5. Age at marriage :
6. Age at first cohabitation :
7. Age at first child birth :

VII. Obstetric History

1. Information about live births

S. No.	*Name*	*Sex*	*Age*	*Was pregnancy period of 9 months complete*		*Physically disable*	*Mentally disable*
				Yes	*No*		
1.							
2.							

Cont'd...

...Cont'd

3.							
4.							
5.							

In case of disability, fill the following information:

8. What is the physical disability?
9. Where was it diagnosed?
10. What is the mental disability?
11. Where was it diagnosed?

2. Information about dead children including still birth

S. No.	*Name*	*Age of the child*	*Sex*	*Cause of death*
1.				
2.				
3.				
4.				

3. Information about abortions

Number of abortions	*Age of mother during each abortion*	*Month of pregnancy*	*Cause*
1.			
2.			
3.			

For the field investigators:

* Which of the following members were present during the interview of Part B?

Informant alone ☐ In-laws ☐ Others ☐

Clinical Examination
(0–5 Years)

Date:

Venue:

[Instruction: Please put a tick mark (v) on the number of the response]

I. Background Characteristics

A. Name of the Child:

B. Village:

1. Mhaswandi 2. Karjule Pathar

3. Kalas Pimpri 4. Kangar

C. Weight:

D. Age:

II. General Clinical Features:

A. Appetite

1. Poor 2. Good 3. Anorexia

B. Thirst

1. Normal 2. Poor

C. Signs of Dehydration:

1. Present 2. Absent

D. Urine

1. Normal 2. Not Normal

E. Stool

1. Regular 2. Constipation

F. Skin

1. Dryness 2. Coarse 3. Pallor

G. Activeness

1. Active/Playing 2. Weak 3. Irritable

4. Moaning 5. Calm/quiet

H. Responsiveness to stimuli

1. Present 2. Absent

I. Any other illness

1. Yes 2. No

Clinical Features of Protein-Energy Malnutrition

A. Physical Growth

1. Normal 2. Average.

3. Poor 4. Retarded

B. Mental Growth

1. Normal 2. Average.

3. Poor 4. Retarded

C. Oedema

1. Present 2. Absent

D. Is Oedema present?

1. Yes 2. No

E. Places where oedema is present

1. Belly 2. Face 3. Legs

F. Muscle wasting

1. Mild 2. Severe 3. Absent

G. Fat wasting

1. Loss of subcutaneous fat 2. Fat retained

H. Diarrhoea

1. Normal 2. Often 3. Absent

I. Skin changes

1. Present 2. Diffuse pigmentation 3. Absent

J. Hair changes

1. Silky 2. Sparse 3. Brown 4. Nil

K. Moon face

1. Often 2. None

III. Nutritional Deficiencies

A. Eye-sight Features

1. Normal 2. Poor 3. Not Applicable

Vitamin A deficiency

1. Present 2. Absent

B. Skin Features: Presence of Pallor/Hypopigmentation

1. Yes 2. No

Vitamin D deficiency

1. Present 2. Absent

C. Anaemia: low Hb (Lower eye-lid)

1. Present 2. Absent

D. Whether Ankle Joint weak?

1. Yes 2. No

E. Oedema

1. Yes 2. No

F. Skin infections

1. Yes 2. No

G. Worms

1. Present 2. Absent

H. Dental caries

1. Present 2. Absent 3. Not Applicable

I. Any serious illness (since birth):

1. Yes 2. No

If yes, details:

Remarks:

Clinician:

Anthropometry of Children
0–5 Years

1. Name of the child : ____________________
2. Name of the mother : ____________________
3. Sex : ____________________
4. Age (in completed years): □□
5. Weight: □□. kg.
6. Height: □□.□ cm.
7. Mid Upper Arm Circumference: □□.□ cm.
8. Head Circumference: □□.□ cm.
9. Chest Circumference: □□.□ cm.
10. If any measurement is not taken, please state the reason:

Monitor's Form

Date of submission : ____________________

Name of Village : ____________________

Total Sample : ____________________

Date of start of data collection : ____________________

Date of end of data collection : ____________________

Write total numbers of the following schedules covered:

Household Schedules	*Mother's Profile*	*Schedules for children*	*Anthropometry*

Name and Signature of Monitor:

Name and Signature of the Receiver of Schedules:

Date of Receipt of Schedules:

Study Consent (Verbal)

I. Introduction

Namaste! My name is *[name]* and I am working with Watershed Organisation Trust (WOTR), Ahmednagar and Research Unit, Pune.

The activity of Watershed development has been going on in rural Ahmednagar for more than a decade. We would like to assess the impact of Watershed development by conducting a study to examine the nutritional status of children of age 0–5 years in rural parts of Ahmednagar. As you have a child of 0–5 years age, your household information is pertinent to our study. Therefore we invite you to participate in our study.

II. About the study

The study is being conducted at Village, Household and Individual levels. Key Informants of the village, a member of the Household and the Mothers of children aged 0 to 5 years are being interviewed. The study also involves a camp for children to assess their nutritional status, by taking their body measurements, called as Anthropometry and a clinical examination by a Physician to detect nutritional deficiencies. The mothers and/or the care takers of the children will be informed about their child's nutritional status in the camp.

III. Study procedure

We would like to have your consent for an interview with you to obtain your household level/village level/individual level information. The topics to be covered in this interview with you are related to land and livestock holding, agriculture allied business, housing, drinking water, fuel, assets, migration, food shortage and loans.

You are free to express your thoughts, opinions, problems and any other matter you would like to share with us.

It will take about an hour at a stretch to complete.

We would also like to have your consent for clicking photographs of your house, the surrounding area, yours, your children and/or your family members and any other relevant matter as a support to our study.

IV. Confidentiality

The information you provide will be kept confidential. The household and individual level data will be identified by assigning a household identification number. The researchers, some senior officials and resource persons of our organisation will have access to the data from the point of view to understand the situation. The findings and the analysis will not relate to the names of any of the study participant.

V. Benefits of the study

There are no direct benefits of the study but at a wider level the study will help us to propose interventions to enhance the nutritional status of children. The study contributes to the society at large by generating scientific knowledge.

VI. Voluntary participation

Participation in the survey is voluntary. In case you have problem with any part of the interview you are free to withdraw your participation. Your withdrawal from the study will not affect your relationship with the organisation and there will be no loss or benefits to which you are otherwise entitled from the point of view of watershed development.

If you have any queries regarding the study or your rights as a study participant, you can contact the Director of our organisation. Please have the contact card.

Have you understood about our study? Do you have any questions?

(The Interviewer must answer any questions and address respondent's concerns).

May I begin with the interview now?

____

Name of Person Obtaining consent

____

Date

References

Agharkar Research Institute (ARI). 1999. *Nutritive values of some Indian Food Preparations*. ARI. Pune.

Bhatia, V. et al., 2007. Malnutrition among Under-Six Children in Chandigarh: Scarcity in Plenty. *Journal of Clinical and Diagnostic Research*, 1(6): pp. 483–487. Available online at http://www.jcdr.net/back_issues.asp?issn=0973–709x&year=2007&month=December&volume=1&issue=6&page=483–487&id=136. Accessed on 6 May 2009

Bhattacharyya, Amiya. 2000 March. Assessment of growth and nutritional status in Indian population. *The Journal of the Anthropological Society of India*. 49(1): pp. 69–102.

Bollom, Michael. 1998. *Impact Indicators: An Alternative Tool for the Evaluation of Watershed Development*. Indo-German Bilateral Project Watershed Management, New Delhi.

Carney, Diana, ed. 1998. Sustainable Rural Livelihoods. Department for International Development. London.

Chauhan, G.S. 2004. Land Degradation Problems: Their Impacts and Remedial Measures for Sustainable Development in Indian

Desert Region in book *Resource Conservation and Food Security-An Indian Experience*. Tapeshwar Singh (ed). Concept Publishing Company. New Delhi.

Deaton, A. and Dreze, J. 2009. Food and Nutrition in India: Facts and Interpretations. *Economic and Political Weekly*, XLIV (7): pp. 42–65.

Deshprabhu, Rashmi. 2001. *Child Development and Nutrition Management*. Book Enclave. Jaipur.

Gopalan, H.N.B. and Saksena, Sumeet, eds., 1999. *Domestic Environment and Health of Women and Children*. United Nations Environment Programme and Tata Energy Research Institute. New Delhi.

Gopaldas, Tara and Seshadri, Subadra, eds., 1987. *Nutrition: Monitoring and Assessment*. Oxford University Press. Delhi.

Gopalan, C., Rama Sastri, B.V. and Balasubramanian S.C. Revised and updated by Narasinga Rao, B.S., Deosthale Y.G. and Pant K.C. 2007. *Nutritive Value of Indian Foods*. National Institute of Nutrition (NIN). Hyderabad. India.

Government of Maharashtra. 1999. Maharashtra Water and Irrigation Commission Report. Vol. I.

Government of Maharashtra. Letter dated 25 May 2006. Declared official list of communities. Copy obtained from Social Welfare Department, Pune.

IGNOU (Indira Gandhi National Open University) School of Continuing education. Year not known. Nutrition for Community, Block 6. IGNOU. New Delhi.

International Institute for Population Sciences (IIPS) and Macro International. 2007. *National Family Health Survey, (NFHS)-3, 2005–2006, India: Key Findings*. IIPS. Mumbai.

International Institute for Population Sciences (IIPS) and Macro International. 2008. *National Family Health Survey, (NFHS)-3, 2005–2006, Maharashtra:* IIPS. Mumbai.

Jaipurkar, Anoop. 2008. Banking on self help. *The Times of India*. Pune, 14 March.

Jazairy, Idriss et al., 1992. *The State of World Rural Poverty*. International Fund for Agricultural Development (IFAD).

Kaur, Ginjinder et al., 2005. Nutritional Status: Anthropometric Perspective of Pre-School Children. *The Anthropologist*. 7(2): pp. 99–103

Lenton, Roberto et al., 2005. Health, Dignity and Development: What Will It Take? UN Millennium Project: Task Force on Water and Sanitation. Available online at http://www.unmillenniumproject.org/documents/WaterComplete-lowres.pdf. Accessed on April 2009.

Levinson, David and Ember Melvin, eds., 1996. *Encyclopaedia of Cultural Anthropology* (Vol. III). Henry Holt and Company. New York.

Maharashtra State Board of Secondary and Higher Secondary Education (MSBSHSE). 2007. *Text Book of Science and Technology*. MSBSHSE. Pune.

NABARD (National Bank for Agriculture and Rural Development). Year not mentioned. Guidelines on participation in Indo German Watershed Development Programme. NABARD. Mumbai.

Narayanmoorthy, A. and Kshirsagar, K.G. 2008. Rapid Impact Evaluation of NWDPRA in Maharashtra, in: Jharwal, S.M. (General editor), Deshpande, R.S., Sharma, V., Malik, R.P.S., Jha, B and Ansari, S.A.,ed. *Glimpses of Indian Agriculture (Macro and Micro Aspects). Volume 2: Micro aspects*, Academic Foundation. New Delhi.

National Academy of Agricultural Sciences, India. 2003. Dichotomy Between Grain Surplus and Widespread Endemic Hunger. Policy Paper 20. Available online at http://www.naasindia.org/Policy%20Papers/pp20.pdf. Accessed on February 2009.

Nutrition Society of India. 2005. Pre Conference workshop on Epidemiological tools in assessment of Nutritional status. National Institute of Nutrition (NIN). Hyderabad.

Pandit, Apoorvaa. 2010. *Watershed Development Inputs and Social Change: Understanding the Changing Culture of Child Nutrition*. Rawat Publications. Jaipur.

Park, K. 2005. *Park's textbook of Preventive and Social Medicine*, 18th Edition. Banarsidas Bhanot Publishers. Jabalpur. India.

Raju, K.V. et al., 2004. *State of the Indian Farmer: A millennium study. Water Resources.* vol. 3. Academic Foundation. New Delhi.

Rajvanshi, A.K. January 2005. Rocket Science for Rural Development. Nimbkar Agricultural Research Institute. http://nariphaltan.virtualave.net/rocketscience.pdf. Accessed on January 2009.

Rao, Hanumantha D. and Vijayaraghavan, K. 1996. Anthropometric assessment of Nutritional Status in Bamji M. et al., ed. *Textbook of Human Nutrition*. Oxford & IBH Publishing Co. Pvt. Ltd. New Delhi.

Samuel, Abraham et al., 2007. Watershed Development in Maharashtra: Present scenario and issues for restructuring the programme. Technical Report. Society for Promoting Participative Eco-System Management (SOPPECOM). Available online at http://www.indiawaterportal.org/tt. Accessed on 30 September 2008.

Subramanian, Sunder. 2002. Water supply and sanitation: India Assessment. Planning Commission, Government of India. Available online at http://planningcommission.nic.in/reports/genrep/wtrsani.pdf. Accessed on January 2009.

Tara Gopaldas and Subhadra Seshadri. 1987. *Nutrition: Monitoring and assessment*. Oxford University press. Delhi.

The Financial Express. 2007. Maharashtra tops the charts in rural sanitation, 30 April. Available online at http://www.financialexpress.com/news/maharashtra-tops-the-charts-in-rural-sanitation/125980/. Accessed on March 2009.

Thimmayamma B.V.S. and Rau Parvati. 1996. Anthropometric assessment of Nutritional Status in Bamji M. et al., ed. *Textbook of Human Nutrition*. Oxford & IBH Publishing Co. Pvt. Ltd. New Delhi.

Watershed Organisation Trust (WOTR). n.d. Watershed Development: An Information Booklet. WOTR. Ahmednagar.

WHO. 1995. Physical status: The use and interpretation of anthropometry (Report of a WHO Expert Committee). WHO Technical Report Series 854. Geneva.

WOTR. 2005. Drop by drop- the story of WOTR. WOTR. Ahmednagar

www.ahmednagar.nic.in. Accessed on 1 September 2008

www.eclac.org/dds/noticias/desafios/4/.../Desafios_Nro2_ing.pdf. Accessed on 20 June 2010.

www.kvk.pravara.com/distprofile.htm. Accessed on June 2009

www.oecd.org/dataoecd/4/24/38148786.pdf. Accessed on May 2009

www.wcd.nic.in/icds.htm accessed on 2 February 2009

www.sampada.org. Accessed on September 2008

Zeigler, Robert. 2006. Bringing Hope Improving Lives. Available online at http://www.irri.org/media/speeches/IRC2006/ IRCPaper.pdf. Accessed on 2009.

Index